VILLAGE ANIMAL HEALTH WORKER TRAINING TEXT and MEDICINE GUIDE

Produced by
Animal Health Training and Consultancy Service
P.O. Box : 419
Ramghat, Pokhara NEPAL
www.ahtcs.org.np

Published by
Christian Veterinary Mission
www.cvm.org

Introduction to Combined Pocket Edition

Christian Veterinary Mission (CVM) is pleased to continue our partnership and cooperation with the Animal Health Training and Consultancy Service (AHTCS). We have been working together since 1980 when Dr. Peter Quesenberry of CVM joined with the Animal Health Improvement Program of the United Mission to Nepal (now AHTCS, an independent organization). This edition has not revised any of the content from the Fourth Edition (available from the author in both English and Nepali) but has produced the textbook and the medicine book in one, smaller pocket edition of the English translation (copyright 2011). We believe that this material, developed and tested for over 30 years by AHTCS in Nepal, is of value to farmers and those who work with them in many contexts outside Nepal. Through provision of this edition, we hope to make this resource more widely available. This book is intended to accompany a training program for village animal health workers and not to stand alone without participatory, hands-on training.

For further Information, please contact:

Author: Animal Health Training and Consultancy Service (AHTCS)
PO Box 419
Ramghat, Pokhara
NEPAL
www.ahtcs.org.np
director@ahtcs.org.np or admin@ahtcs.org.np
Tel. +977-61-526455/530952
Fax. +977-61-535807

Publisher: Christian Veterinary Mission
vetbooks@cvm.org
www.cvm.org

IMPORTANT ! LIST OF REPORTABLE DISEASES

THE FOLLOWING DISEASES SHOULD BE REPORTED IMMEDIATELY TO YOUR NEAREST GOVERNMENT VETERINARY POST

1) RINDERPEST AND PPR

Rinderpest is a disease that affects cattle, buffalo, sheep and goats. PPR affects goats and sheep. Both are very serious diseases that make many animals sick at the same time and may kill many animals if not rapidly controlled by vaccination.

The signs are:

- Discharges from the eyes and possibly the nose
- Diarrhoea
- Death
- You may see small lesions in the mouth and crusting on the vulva
- Additionally, in PPR, many animals develop a cough due to pneumonia

IF YOU SEE THIS COMBINATION OF SYMPTOMS IT IS ESSENTIAL THAT YOU REPORT THEM IMMEDIATELY TO THE LOCAL VETERINARY AUTHORITIES (SEE PAGE 197 OF THIS MANUAL).

The other diseases or situations which you must report as soon as possible to the local veterinary authorities are:

2) RABIES
3) BLACK QUARTER
4) HAEMORRHAGIC SEPTICAEMIA
5) RHANIKHET (NEWCASTLE DISEASE)
6) ANY DISEASE WHICH KILLS 5 OR MORE ANIMALS IN ONE WEEK IN THE SAME VILLAGE
7) ANY DISEASE WHICH MAKES 10 OR MORE ANIMALS SICK IN ONE WEEK IN THE SAME VILLAGE

They will then be able to help you by investigating the problem and arranging necessary measures to prevent the disease spreading.

Introduction to First Edition

Acknowledgments:

This book is the product of the hard work of the staff of the Animal Health Improvement Training Programme. Both Nepalese and Ex-patriate staff have been working together for over ten years to produce this book. We are especially grateful to the following people:

Dr. Alison Craven, Veterinarian; Dr. Peter Quesenberry, Veterinarian; Marteen van den Dool, Animal Husbandry Specialist; William van Dis, Animal Husbandry Specialist; Dr. Barbara Fargher, Veterinarian; Dhananjaya Paudel, Animal Heath Development Officer; Dr. Wade Bradshaw, Veterinarian; Parashuram K. C., AHITP Farm Manager; Raju Chitrakar, Illustrator and Layout Artist; Usha Gurung, Secretary; Beena Tulachan, Secretary

Rural Development Centre, 1989 (2046 BS)

Introduction to Second Edition

In writing this second edition of the Village Animal Health Worker Textbook, we wanted to concentrate on what were the best ways to keep farmers' animals healthy. We took suggestions and communication from HMG and NGO personnel as well as from our own AHITP staff. We also considered comments taken from our VAHW trainees. We think that the result is a significant improvement over the first edition. The following people contributed much to the publication of this book:

Dhananjaya Paudel, Animal Development Officer; Narayan Dutta Ojha, Animal Development Officer; Goma Kumari Shrestha, Animal Health Technical Facilitator; Henk-Peter Diikema, Animal Development Specialist; Dr. Karen Stoufer, Veterinarian; Parashuram K. C., Farm Manager; Raju Chitrakar, Illustrator and Layout Artist; Usha Gurung and Beena Tutachan, Secretaries; Gopal Thapa and Indra K.C., Animal Health Technical Facilitators; Jhalak Kanal, Language Correction

Rural Development Centre, 1991 (2047 B.S.)

Introduction to the Third Edition

Over the 17 years we have been training VAHWs, we have obtained input from many different organisations and over 1000 trainees in the Animal Health Improvement Training Programme. Extensive field-testing has shown this book to be very practical and useful. Much effort has been invested over the past year to make this third edition even better. In this edition, all the topics have been extensively revised and many new ones added so that this curriculum will prepare the VAHW to successfully complete the skill test for VAHWs administered by the Council for Technical Education and Vocational Training (CTEVT). Special thanks to all those who worked hard to bring about this new edition:

AHITP Staff

Dr. Dilli Ram Bhandari, Dr. Shanti Upadhaya, Dr. Robert Hott, Dr. Horn Bahadur Basnet, Mr. Narayan Dutta Ojha, Mrs. Parbati Gurung, Ms. Indira K. C., Mr. Kedar Ghimire, Mr. Indra Bahadur K. C., Mrs. Nim Kumari Khadka, Mr. Tulasi Jung Dahal

Secretarial Assistance

Ms. Mamta Rana, Ms. Manmaya Gurung, Mrs. Beena Tulachan
John Ferchak, Director, RDC, (May 1998)

Introduction to Fourth Edition

This book is successfully revised to fill a long standing need of a text book for Village Animal Health Workers working in different field situations. It gives basic necessary ideas about the various topics dealt within the scope of workers. The contributors have tried to revise and add the topic in as concise a manner as possible, laying stress on fundamentals. The topics are dealt with in a simple way, avoiding complicated explanatory details. The VAHW trainees are required to stick to this text book. Text in this book has been kept to a minimum and the many drawings visualize the subject matter.

The present English version was revised following the requests of Volunteers in Irish Veterinary Assistance (VIVA), Ireland and BOTHAR Ireland. In this context, we are very grateful to Dr. Brendan Mimnagh of VIVA and Peter Ireton of BOTHAR for their continuous support and inspiration. We are also grateful to Mr. Umesh Lama of UMN for his invaluable support to come up with this edition.

In revising this text book, the comments and contributions of the following personnel are gratefully acknowledged: Dr. Dirga Kumar Lamichhane, Dr. Susan Griffin, Dr. Resham Prasad Acharya, Dr. Sabina Shrestha, Dr. Paras Thapa, Dr. Sadhana Shrestha, Dr. Jeewan Thapa, Dr. Bed Bahadur Chhetri, Dr. Dilip Prasad Bhandari and Mr. Shiba Kumar Pradhan. Secretarial Assistance : Magan Shrestha, Rajan Khadka

Dr. Dirga Kumar Lamichhane, Executive Director, AHTCS (August, 2007)

Table of Contents

Introduction

Examination

Medicine

Table of Contents

Animal Care

Parasites

Table of Contents

Table of Contents

Respiratory Diseases

Common Diseases

Zoonotic Diseases

Table of Contents

Responsibilities of Village Animal Health Worker

The ultimate goal of Village Animal Health Worker (VAHW) training is to produce VAHWs with the ability to treat and control animal diseases and problems at village level.

VAHW role and responsibilities are as follows:

1. Examine sick animals at the request of farmers and choose the right treatment to help these animals get better. They should follow up these animals to see the results of the treatment.

2. To coordinate with village authorities and farmers to organize vaccination campaigns in time to prevent disease outbreaks.

3. To record livestock data in the village: births, deaths, diseases and sales. By recording this data and the dates they can assist villagers and veterinary staff to solve problems and increase production.

4. Report disease outbreaks to the District Livestock Service Office. During outbreaks they should assist veterinary staff in sending samples to the veterinarian's office.

5. To advise villagers about basic livestock husbandry and management techniques (nutrition, breeding, housing, etc.). They should incorporate these techniques themselves and demonstrate them to other farmers.

Restraint of Animals

Why?

- To examine the animal properly.
- To give injections and other medicines.
- For the animal's safety as well as your own safety.

Methods of Restraint:

1. Tying the head and front feet:

The two neck ropes prevent the animal from moving around the stake in front. Tying the front feet prevents the animal from jumping up or moving around. This method can be used both for examination and for giving injections.

Note: Holding the tail straight up, over animal's back will help prevent the animal from kicking.

2. Tying up a front leg:
Another method of restraint is to tie the rope to one front leg just above the hoof. Pass the rope around the body of the animal as shown and ask a helper to hold the end. Pulling on the rope will raise the front leg so the animal stands on three legs. This will prevent the animal from moving.

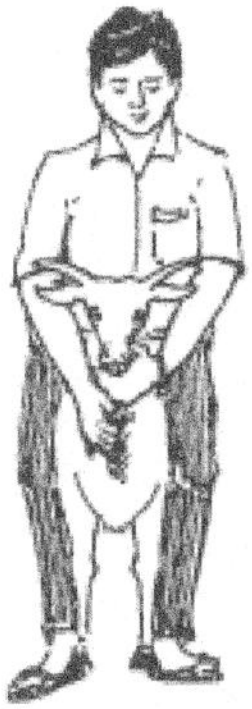

3. Holding between the legs:

Goats and sheep can be restrained by trapping them between the legs of your helper.

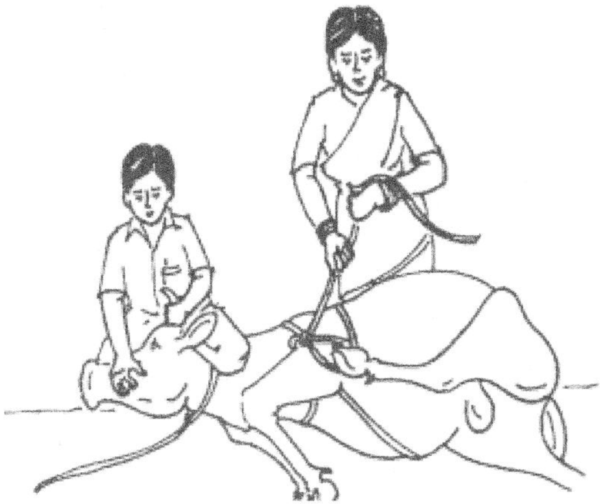

4. Casting (Laying Down):

Animals should be cast (laid down) for castration, minor surgery, and for treatment of foot wounds. There are many methods of casting. The method used should be chosen based on the type of animal. Methods used for bullocks are not good for milk cows and buffaloes. A soft cotton rope should be used and always placed in front of the udder.

Materials needed for casting:

- Strong wooden stake about 2½ meters long.
- Long cotton rope about 10 meters long
- Some helpers and a neck rope.

1. Stake (wooden pole)

The stake should be buried securely in the ground.
The neck rope should be at ground level.

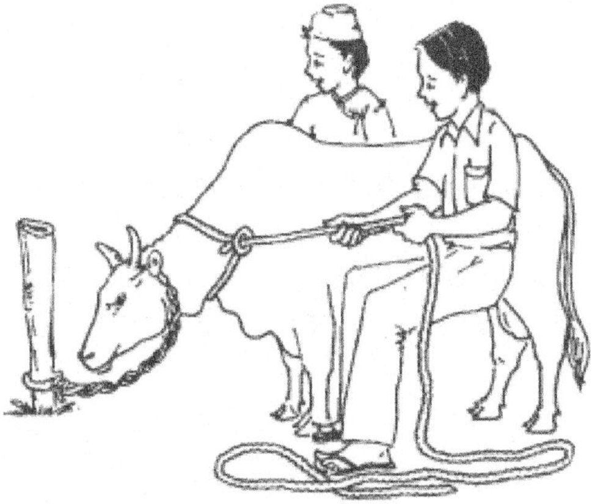

2. Long rope

The casting rope should be tied around the neck with a non-slip knot. Always be sure the casting rope is behind the neck rope.

How to make a non-slip (bowline) knot:

- There are many methods to tie a non-slip knot
- This is a good one:

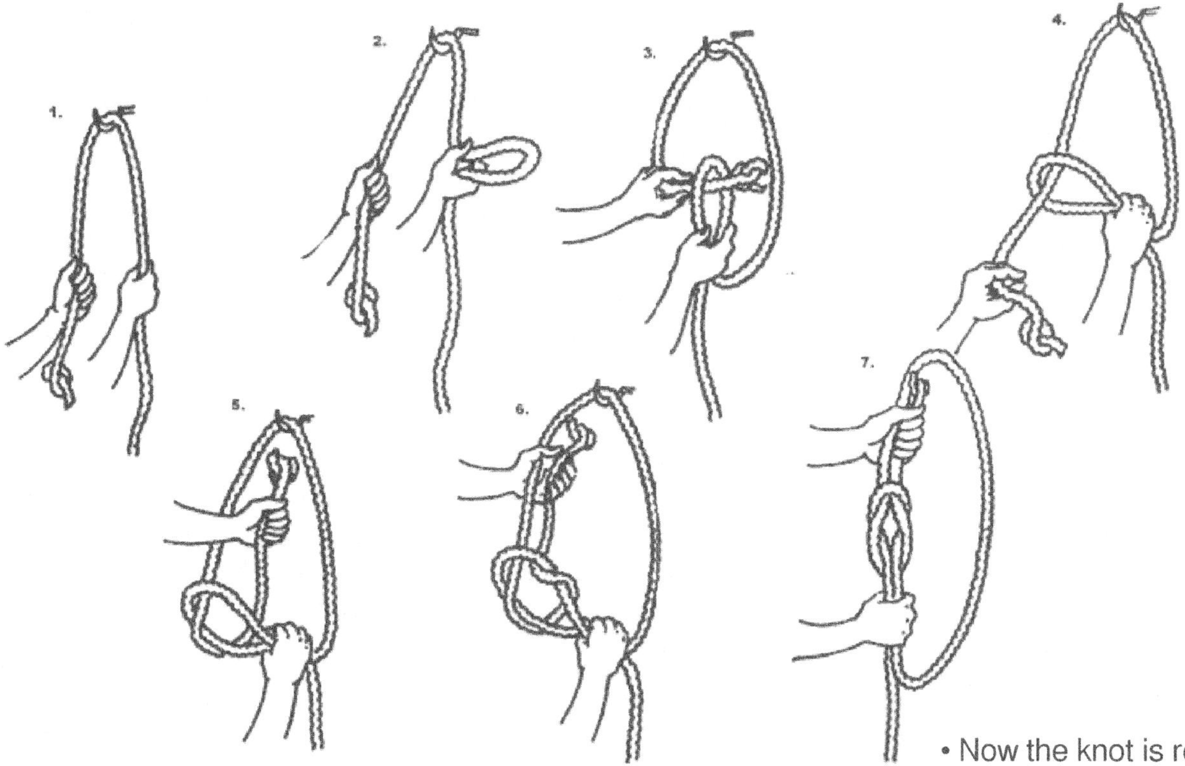

- Now the knot is ready.

3. One person should stand on the left side of the animal, and one should stand on the right side.

4. Make a loop behind the front legs by passing the rope over the top to the right side, then pass the rope underneath and back to the left side.

5. After passing the rope underneath to the person on the left, tighten it behind the hump.

6. Pass the rope again over the animal from left to right.

7. The rope on the left side should be pulled around in front of the hip bones, made into a loop, and pulled tight.

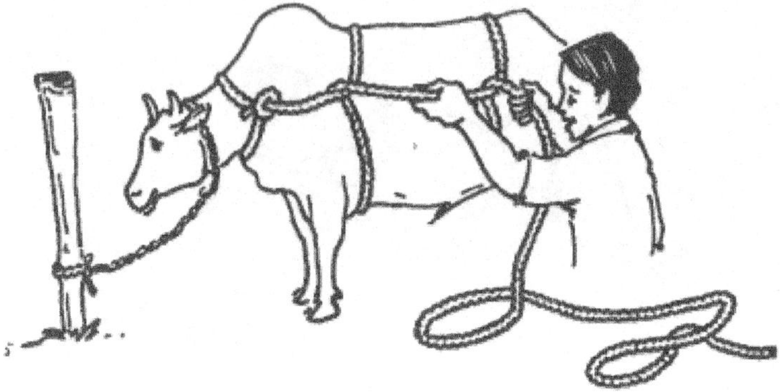

8. One or two strong helpers should pull the rope straight backwards.

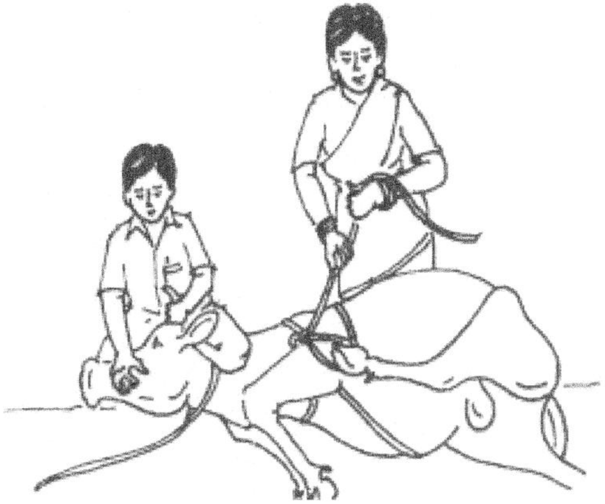

9. One helper should be ready to hold the head down on the ground.

10. After the animal is cast, the long rope should be wrapped around the upper back leg and brought through the neck loop. This will keep the animal from getting up. The rope should be pulled tight and tied securely.

The casting method shown at right is good for milk cows, milk buffaloes, and horses. One short rope is used to tie the front feet together. The long rope is tied to the back feet and looped around the short rope as shown. Casting is done by pulling backwards on the long rope and tying the hind feet as shown above.

5. Restraint in a crate (head gate)

- Good for examination.
- Good for giving injections and drenching.

Appropriate place to build a crate:
- Should be in a central location for the village.
- Should be in a place where it is easy for farmers to bring their animals.
- Should be built in a place where it can be used for a long time.

Materials needed to build a crate:
Bamboo or wood, rope, knife or axe, nuts and bolts, iron rod to dig holes, measuring tape, and drill.

Size of crate:
- 2 meters long
- 1 meter wide
- 1.5 meters high and ½ meter below the ground level

Restraint of Chickens

What is it?

Restraint of chickens and poultry is done to prevent them from moving while we examine them, give injections, cut the beak, or transport them from place to place.

1. Restraint by folding the wings

- Hold the bird with 2 hands

- Hold each wing separately in each hand

- Place one wing behind the other and twist once again.

2. Other methods of chicken restraint:

a) Large birds can be controlled by holding onto their legs and hanging them upside-down.

b) Small chicks can be controlled just by placing your hands over their wings.

Be careful: Sometimes chickens may die if kept restrained for too long!

Health & Diseases

What is Health & Disease?

Health is a body condition which keeps an animal as productive as possible.

Disease is a body condition which causes an animal to be less productive than normal.

What is a Healthy Animal?

A healthy animal is one that is

- Active and Alert
- As productive as possible
- Gives benefit to its owner

What is a Diseased Animal?

A diseased animal is one that is

- Showing some disorder
- Not as productive as normal
- Giving less benefit to its owner than normal

What is an Infectious Disease?

A disease caused by living organisms (Bacteria, Virus, Protozoa and Fungi)

Disease caused by Bacteria	• Hemorrhagic Septicaemia (HS) • Black Quarter (BQ) • Anthrax • Brucellosis • Enterotoxaemia • Foot Rot • Tetanus • Tuberculosis • Mastitis • Pneumonia etc.
Disease caused by Internal Parasites	• Liver Fluke • Round Worms • Gid (Tape Worms Larvae)
Disease caused by External Parasites	• Lice • Ticks • Fleas • Mange Mites • Maggots in Wound

Disease caused by Virus	• FMD (Foot and Mouth Disease) • Rinder Pest • Ephemeral Fever • PPR (Peste-Des-Petits-Ruminants) • ORF • Rabies • Swine Fever
Disease caused by Fungi	• Ringworm • Fungal Toxicosis (Mycotoxicosis)
Disease caused by Protozoa	• Babesiosis

What is a Contagious Disease?
It is an Infectious disease that spreads directly from animal to animal.

Contagious Disease
- HS
- Anthrax
- Rabies
- FMD
- Rinderpest
- PPR
- ORF
- Swine Fever
- Mange Mite

What is a Non-Contagious Disease?
It is an Infectious disease that does not spread directly.

Non-Contagious Disease
- Tetanus
- Ephemeral Fever
- Liver Fluke
- Gid
- Babesiosis

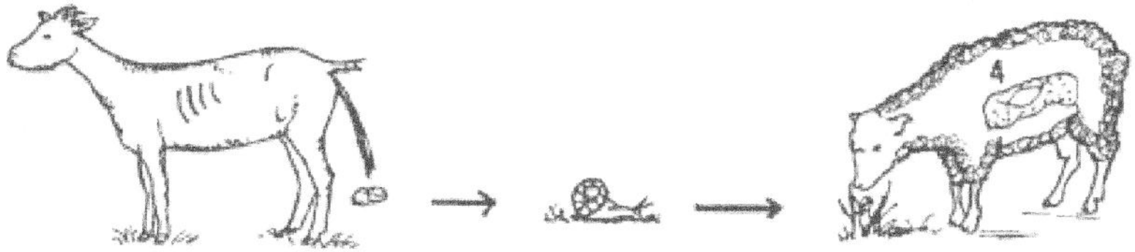

Note :
- All infectious diseases are not contagious diseases
- All contagious diseases can be infectious

What are the Causes of Disease?

- **Infectious Causes:** Bacteria, Virus, Protozoa, Parasites, Fungi
- **Non-Infectious Causes:** Injury, Poisonous Plant/Chemical, Malnutrition, Poor Sanitation

What is Acute Disease?
Acute Disease begins rapidly (quickly) and does not last for a long time.

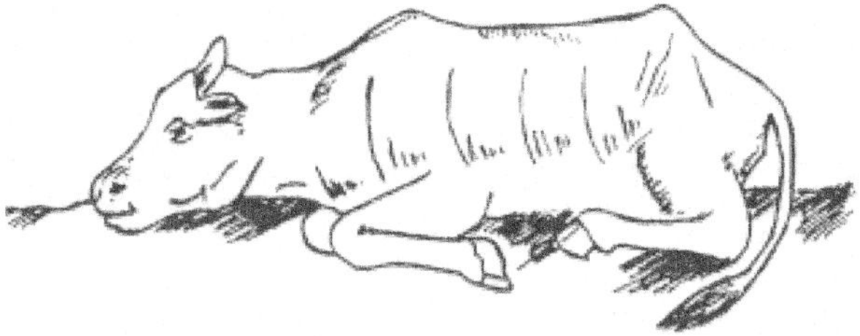

What is Chronic Disease?
Chronic disease begins slowly and lasts for a long time.

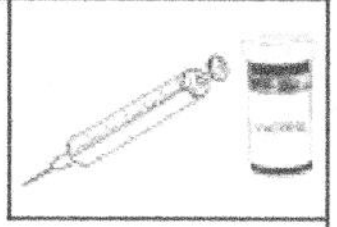

Resistance & Immunity

What is Resistance?

It is the ability to avoid disease or sickness.

This animal will be able to resist disease

Better than this animal

What is Immunity?

It is the ability to protect itself against a particular infectious organism.

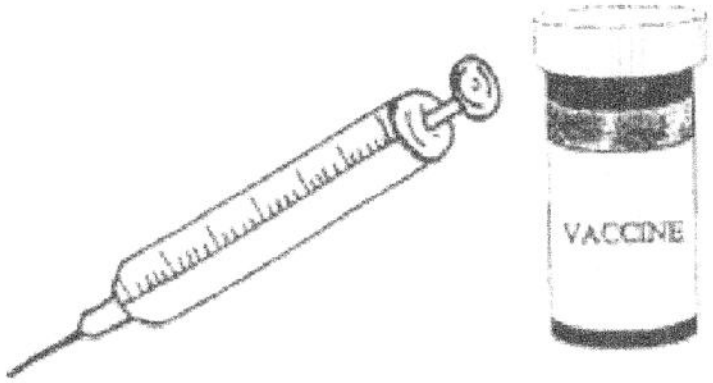

Active Immunity

Use of vaccines gives active immunity to animals.

Passive Immunity

Feeding colostrum to new born animals gives passive immunity

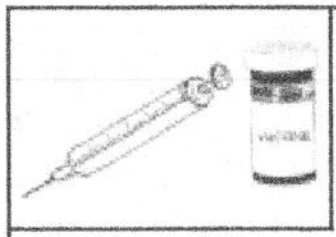

Basic Principle of Vaccine Use

Many types of vaccines provide immunity of varying duration against different diseases. Always follow instructions and read the precautions provided with the vaccine. Remember the following general principles:

- Obtain vaccine from a reputable source. Poor quality vaccines can kill, and mishandled vaccines can lose their effectiveness.
- A vaccine is used to prevent disease, not to treat it. Once the animal is sick, it is too late to vaccinate it against that disease. Vaccinating sick animals may make the disease worse. Do not vaccinate sick animals.

Most vaccines provide protection after one or two weeks of vaccination. If the animal already has the organism in its body when the vaccine is given, or if the animal is exposed to the organism within one or two weeks after vaccination, animals may still become ill. Some vaccines require two injections given three to four weeks apart to provide good protection. The animal is therefore not protected until a few weeks after the second vaccination.

Maintaining Cold Chain

The process of keeping the vaccine at the correct temperature from the moment it is manufactured, to the moment it is administered, is called the "Cold Chain".

For the vaccines to work properly and effectively, cold chain needs to be maintained. Most vaccines are kept at 17°F (-8°C). Some vaccines when stored for longer periods of time (i.e. more than one month) should be frozen. Though some vaccines can become ineffective if frozen, other vaccines are not sensitive to freezing, but lose their effectiveness if they become warm.

For many vaccines if the cold chain is broken the vaccine is ruined, or at least less effective.

There are few vaccines which do not need refrigeration. These vaccines are called heat stable vaccines.

Examination of Sick Animals

There are three parts:
Part 1: Collecting information about the sick animal (history taking)
Part 2: Examination of the sick animal.
Part 3: Examination of the environment.

Why is this important?

- To properly diagnose the disease.
- To give proper treatment.
- To give proper advice to owner.

Part 1: Collecting information about the sick animal (taking the medical history)

Much can be learned by asking questions about the sick animal. If you can see the animal in its own shed and talk with the owner directly, this is even more helpful.

How do we take a medical history?

There are 11 questions we must ask the owner:

1) **Why have you called?**

(Or)

What is the problem?

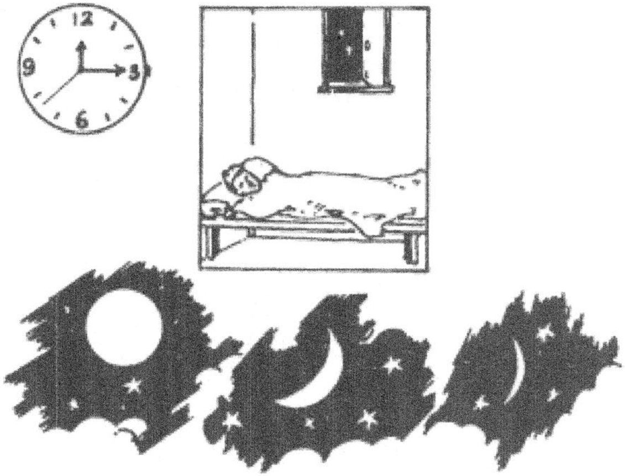

2) How long has it been sick?

- How many hours?
- How many days?
- How many weeks?
- How many months?

3) Which part of the body is affected? What is noticeable?

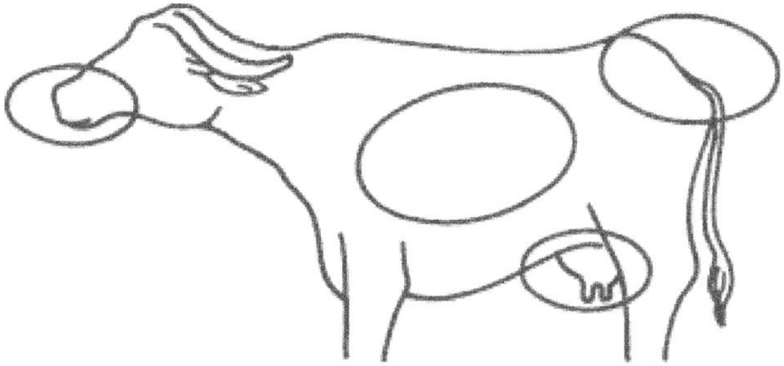

4) Do you raise any other animals?

Are your other animals sick?

Are your neighbors' animals sick?

5) How old is the animal?

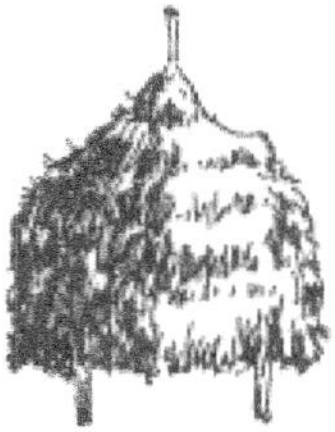

6. Is the animal eating and drinking?

What kind of food is it eating?

7. How are the urine and faeces?

- Constipation or diarrhoea?
- Blood in the urine?
- Blood in the faeces?

8. Is the animal ruminating (chewing its cud)?

- Cattle, buffalo, sheep, and goats are ruminant animals.

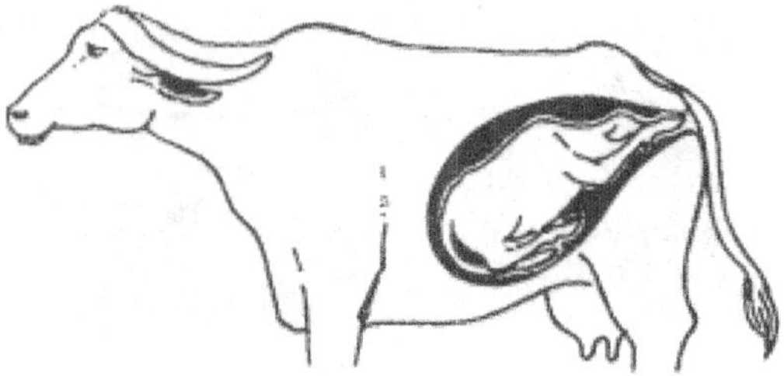

9. Is the animal pregnant?

- If so, when will the baby be born?

10. Is the milk production normal?

- Is the color and smell of the milk normal?
- Does the milk coagulate (clump) when boiled?
- Is the calf nursing normally?

11. Has any other treatment been given?

- What?
- When?
- By whom?
- Was it successful?

Part 2: Examination of the Environment/Surroundings

It consists of two parts

1. Outdoor Environment
2. Indoor Environment

1. Outdoor Environment

A VAHW should always check carefully for the following aspects:

a. Soil and Pasture

- In what type of pasture did the animal graze?
- Is the pasture damp and marshy? (This may cause liver fluke infestation, mosquito borne diseases etc.)
- Does the pasture contain any poisonous plants?
- Does it contain any garbage, industrial wastes, or agricultural pesticides (organophosphates)?

b. Drinking Water Supply

- Is the water source clean and fresh?
- Is it contaminated with sewage disposal or industrial wastes?

c. Weather

- Was the animal exposed to harsh weather conditions like extreme hot, cold or wind?

2. Indoor Environment

a. Housing / Sheds / Pens

- Does it have adequate ventilation?
  (Lack of ventilation leads to respiratory problems)
- Is the hygiene in the shed always maintained?
- Does it have adequate space to keep the animals?
- Are the feeding and watering pans kept clean?

b. Feeds

- Do the animals have a properly balanced feed?
- Is the feed mouldy or contain any poisonous materials or toxins?
- Is there access to a mineral block?

c. Water

- Is there access to plenty of clean water?

Part 3: Examination of the sick animal

a) Examination from a distance:

Before doing a close-up examination of each part of an animal, one should always observe the animal from a distance. Observing from a distance can provide vital information about the animal's behaviour and general condition.

What can we observe from a distance?

- Is the animal's breathing normal?
- Is the animal thin or fat?
- Can the animal walk normally?
- Is the animal active or depressed?
- Groans?
- Scratching?
- Looking at area of pain/discomfort?
- Looking at any particular area of body?

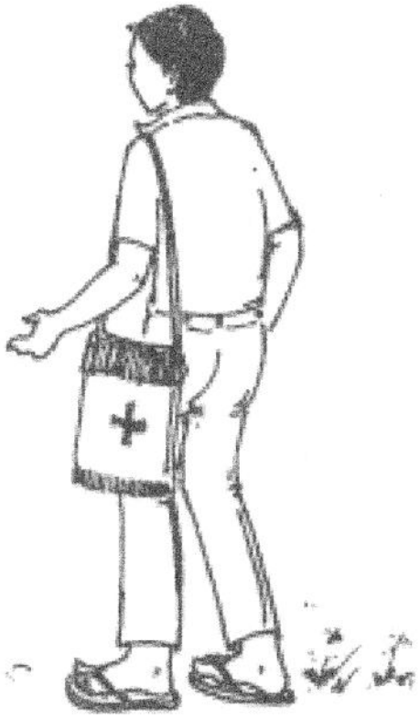

b) Close-up examination:

First of all the animal must be adequately restrained. A crate can be useful for this.

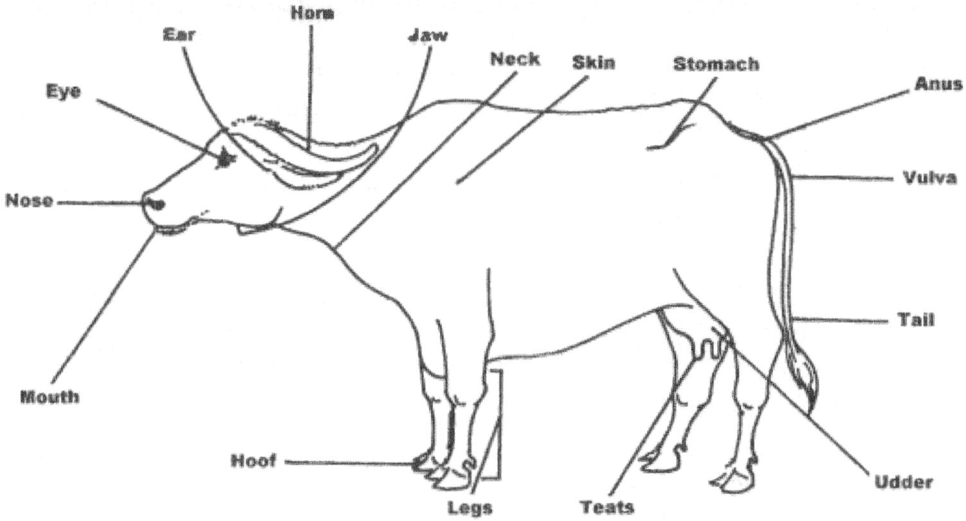

Order of examination:

Mouth, nose, eyes, horns, throat, neck, skin, stomach, anus, vulva, tail, udder, teats, legs, hooves.

Examination of Faeces

Collection:

- From a fresh sample on the ground.
- By placing a hand (in large animals) or a finger (in small animals) into the rectum.

Observations:

- Amount - Normal
  - None (not eating, blockage)
  - Lots (diarrhoea)
- Colour - Brown-Normal
  - Red in parts/yellow (diarrhoea/digestive disorder)
  - Black (Stomach ulcers or bleeding in the digestive system)
- Consistency - Solid and Soft - Normal
  - Solid and hard/dry (dehydrated constipation)
  - Very soft/watery (diarrhoea)
- Odour - Normal
  - Very bad (digestive disorder)

A sample may be taken for lab examination at this time.

Examination of Urine

Collection:

- The owner may collect a sample in a clean container when the animal is urinating. Then they can bring the sample to you.
- You can persuade the animal to urinate by gently rubbing a handful of dry straw against its vulva (for a female) or penis (for a male). The tail should be left in its normal position.

Observations:

- Amount - Normal
  - None (could be technique of collecting, blockage, too painful i.e. an infection)
  - Lots (dribbling urine - muscle damage after a hard birth)
- Colour - Normal
  - Red (blood in urine from dehydration, stones, disease, or after birthing)
  - Cloudy (infection - bladder/kidney)

A sample may be taken for lab examination at this time.

Examination of Milk

Collection :

- Milk a small amount of milk, into your hand or a container, from one teat (after observation repeat for other teats).

Observations:

- Amount - Normal
  - None (blockage, infection, animal not lactating)
- Colour - Normal
  - Brown/Yellow (may be infection or colostrum)
  - Red (generally colostrum, may be infection)
- Consistency - Normal
  - Clotted/lumpy (mastitis)
  - Watery (mastitis)
- Odour - Normal
  - Bad (mastitis)
  - Bad when boiled (mastitis)
  - Boil if unsure that the milk is bad.

A sample may be taken for lab examination at this time.

Possible abnormalities of various body parts:

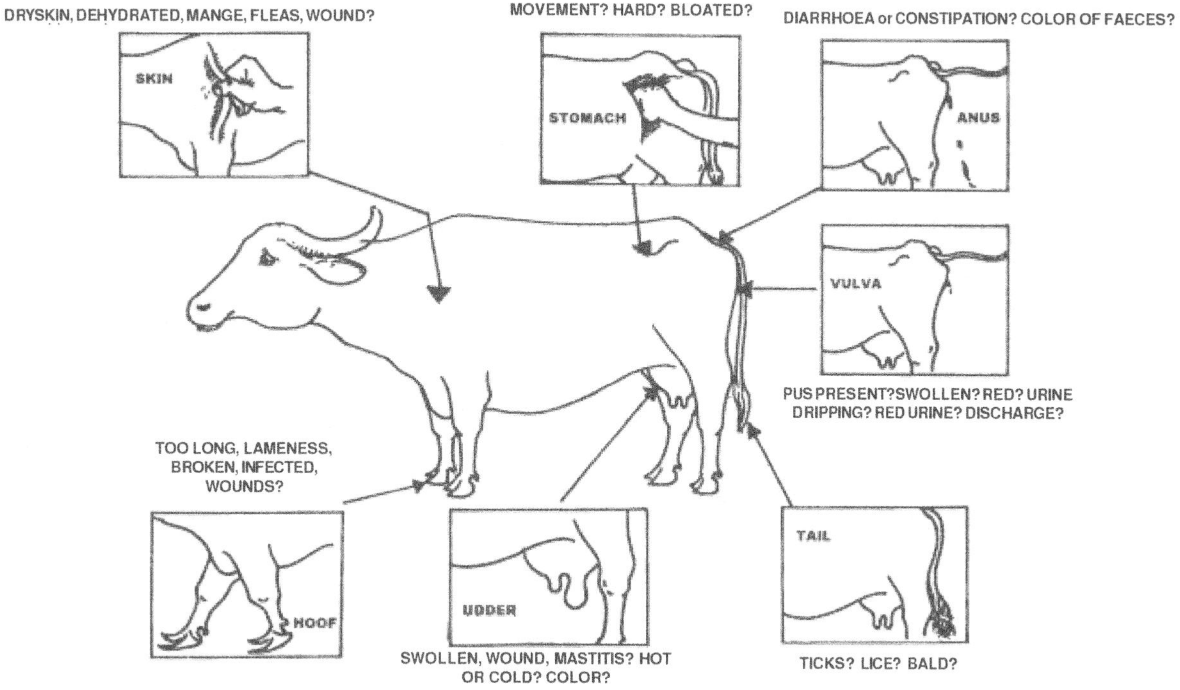

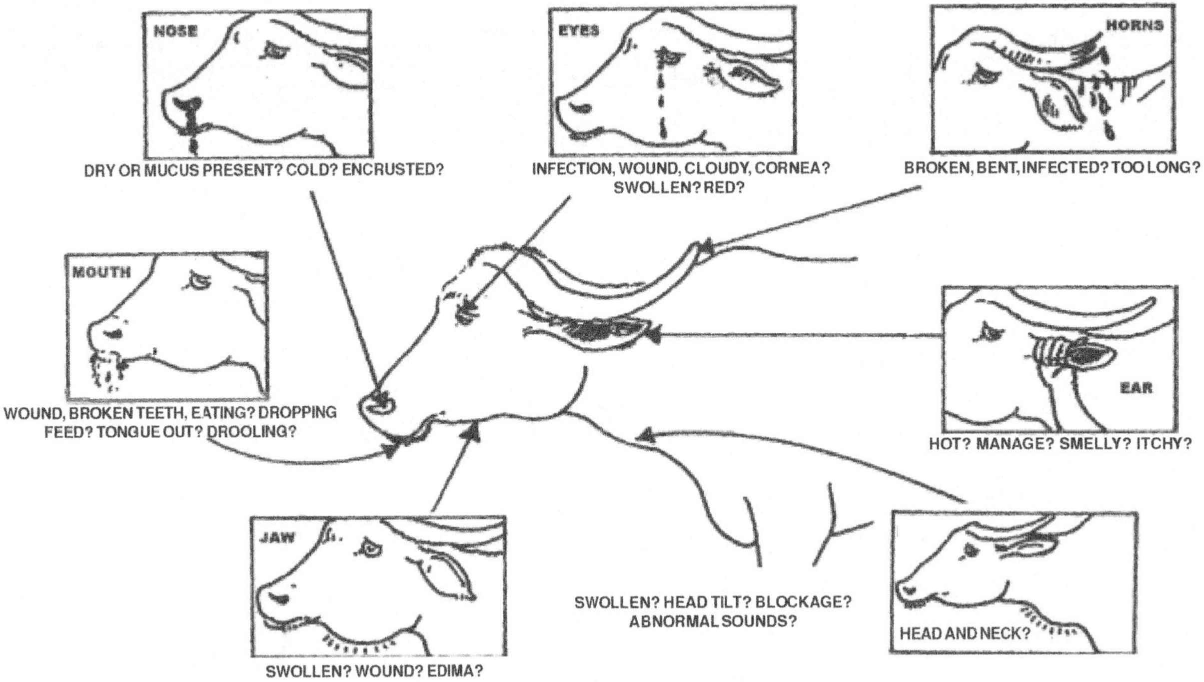
NOSE
DRY OR MUCUS PRESENT? COLD? ENCRUSTED?
EYES
INFECTION, WOUND, CLOUDY, CORNEA? SWOLLEN? RED?
HORNS
BROKEN, BENT, INFECTED? TOO LONG?
MOUTH
WOUND, BROKEN TEETH, EATING? DROPPING FEED? TONGUE OUT? DROOLING?
EAR
HOT? MANAGE? SMELLY? ITCHY?
JAW
SWOLLEN? WOUND? EDIMA?
SWOLLEN? HEAD TILT? BLOCKAGE? ABNORMAL SOUNDS?
HEAD AND NECK?

Thermometer

What is it?

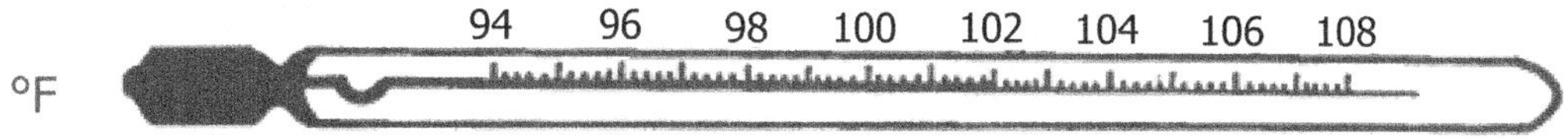

This is an instrument used to measure the body temperature of an animal. Your thermometer may use the Fahrenheit (F) or Celsius (C) scale.

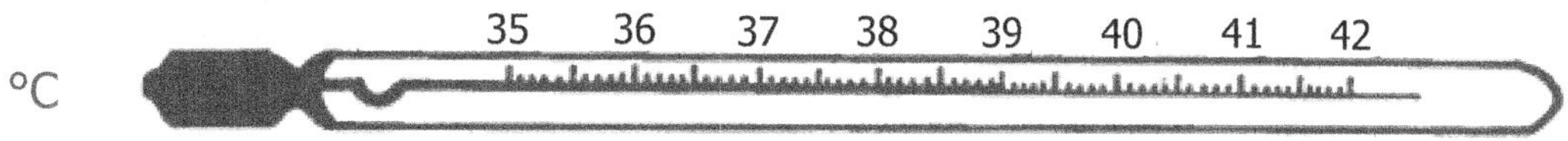

The thermometer has 2 parts.

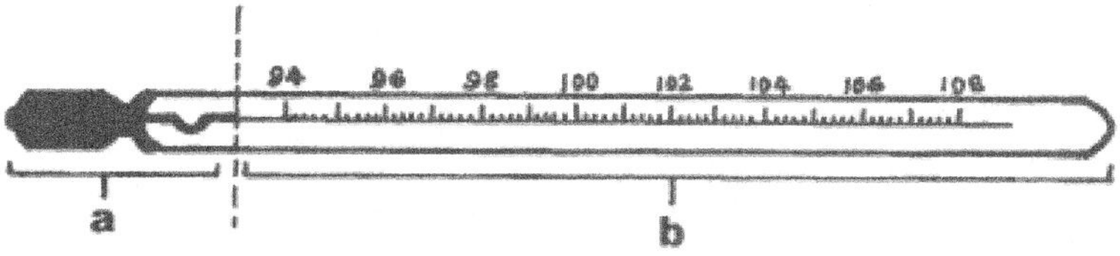

a) The part with the mercury.

b) The part with the degree graduations marked on it.

On the thermometer we can easily read 94, 96, 98, 100, 102, 104, 106, and 108 degrees. (Or 35-42 on a Celsius thermometer.)

But the numbers 95, 97,99,101,103,105, and 107 are not written on the thermometer. These numbers are indicated by a long line between 2 numbers. For example, between 94 and 96 is 95. In the same way, 97 is between 96 and 98, and so on up through 107.

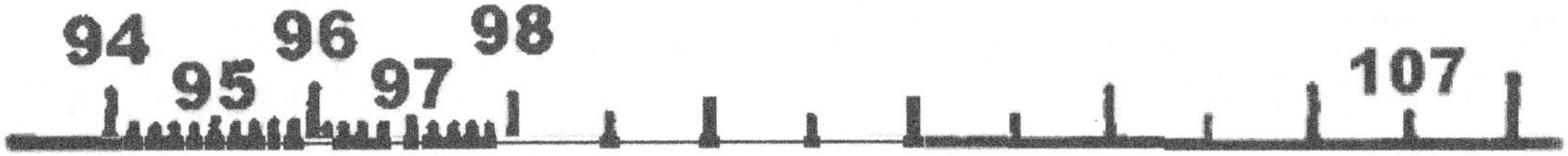

Why do we use a thermometer?

At the end of the examination we use the thermometer to measure the animal's body temperature. Changes in body temperature can be a sign of disease.

How do we use the thermometer?

1. First we must bring down the mercury to the end of the thermometer. Holding the thermometer securely by the end with the degree marks, shake the thermometer to bring down the mercury.

2. Make the mercury end slippery. Spitting on the thermometer can be used to make it slippery. This will allow you to insert the thermometer easily.

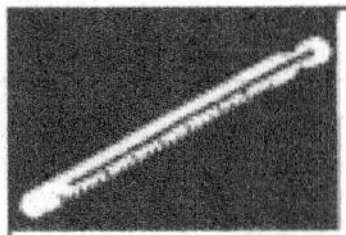

3. Insert the mercury end of the thermometer gently inside the anus. Two thirds of the thermometer must be inserted, with 1/3 outside.
4. The mercury end of the thermometer must touch the rectal muscle. If the thermometer is inserted straight it will likely be buried in fecal material. Therefore, place the thermometer to one side.
5. The thermometer must remain inside the animal for 2 minutes. Remove the thermometer after two minutes. Clean it by wiping gently on the animal's hair or with some cotton. The animal's body heat causes the mercury to rise. The mercury will stop at a certain point. The point where the mercury stops rising indicates the animal's body temperature.
6. Read the thermometer by holding it about 30 cm away at eye level. Roll the thermometer slowly to see the mercury. Sometimes it can be helpful to put your hand behind the thermometer.

Things to remember:

- Keep the thermometer away from heat. For example, do not store it beside the fire or in a tin box out in the sun.
- Do not let children play with the thermometer. It is easy to break it and the mercury is poisonous.

Respiration & Pulse Rate

Respiration

Respiration means breathing. The number of times an animal breathes per minute may change according to environmental or infectious conditions.

In the following conditions, the respiratory rate may increase:

- Fever
- Infection in lungs
- Infection in heart
- Anemia
- Blockage of respiratory tract
- Abdominal pain

Remember the increase in respiration rate in the following conditions is NORMAL

1. During excitement or restlessness
2. During stress
3. Exercise or work for a long time
4. Animals exposed to high environmental temperatures and humidity
5. Advanced pregnancy
6. Animals not kept in well ventilated rooms

How to determine the Respiration Rate?

1. By holding the back side of your hand in front of the nostrils and counting the number of times per minute that the air comes out of the nostrils and hits your hand.
2. By holding a paper in front of the nostrils and counting the movements.
3. By observing the movement of the ribs.

Pulse

It is the number of heart beats per minute. It tells the condition of heart. Pulse rate may be strong or weak, rapid or irregular.

How to Measure Pulse Rate?

Place a tip of your finger against an artery and count the number of pulsations per minute.

Note:

1. A rapid pulse rate indicates fever, fear, pain, heart problems, prolonged exercise or exposure to hot environment.
2. A weak, irregular pulse rate indicates that the heart is not working properly or the animal is in a state of 'shock'.
3. Always keep the animal at REST with no external disturbances while measuring the pulse rate.

Medical Records

What is a medical record?
Written details of a sick animal including the history, treatment, and outcome of the case is called a *medical record* or a *written case history.*

Why do we keep written medical records?

- We cannot remember the details of every case, so we must write them down.
- To serve as a record of our work. We can show it to the local authorities and to the instructor at follow-up time.
- To remind us of the date to give medicine for internal parasites.
- To help keep inventory of our medicine.
- To remember which services were given.
- To find out which diseases are most common.
- To find out what time of the year most problems occur.

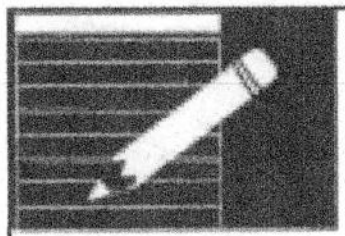

How do we keep written medical records?

After examining the animal, record only the abnormal findings. We do not need to record the findings that are normal. For example, we do not need to record that the animal is eating normally, but we always record the temperature and the stomach movements.

Temperature: Use the thermometer to measure the temperature. Record the exact number of degrees in your record notebook.

Movement of the stomach (rumen): Record the number of movements the stomach makes in 2 minutes by placing your hand on the left flank. Normally, it is three movements in 2 minutes.

Example of Case Record Form:

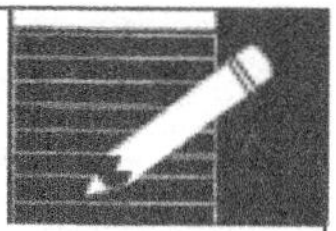

Date: 1/1/98	What was the problem? What was found on examination?	Medicine used	Dose given	Doses given in one day	How many days of treatment?	Advice to owner:
Name of owner: Kancho Magar Address: Hajardanda 8 Magar Tole Animal: Buffalo Age: 7 years	<u>History:</u> No appetite for 2 days, diarrhoea <u>Near exam:</u> conjunctiva white, dehydrated, 2 rumen movements in 2 minutes <u>Temperature:</u> 102° <u>Examination from far :</u> Lean and thin	Vitamin B injection, Distodin	10ml 3 tablets	1 dose 1 dose	3 days 1 day	Give fresh grass and feed, give plenty of water, keep shed clean and dry. Give LF medicine every 6 months

- Use one section for each animal in the record form
- Many animals' records can be kept in one section of the form if they are all being treated for the same problem.

Forms of Medicine and Their Use

Part I: Forms of Medicine

Medicines come in many forms: Liquids, powders, tablets, ointments, and injections.

1. Liquids

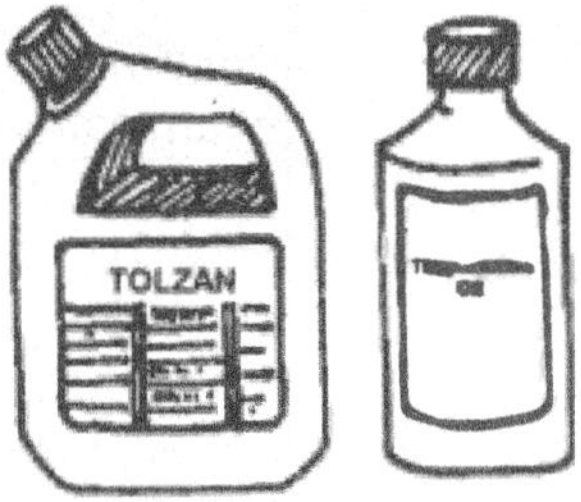

These vary from watery to as thick as yogurt.
Examples: CTC, Tolzan®, turpentine oil.

2. Powders

These vary from fine flour-like powder to rough crystals like salt. Their color, smell, and texture also vary.
Examples: Neblon®, Timpol®, Magnesium sulphate

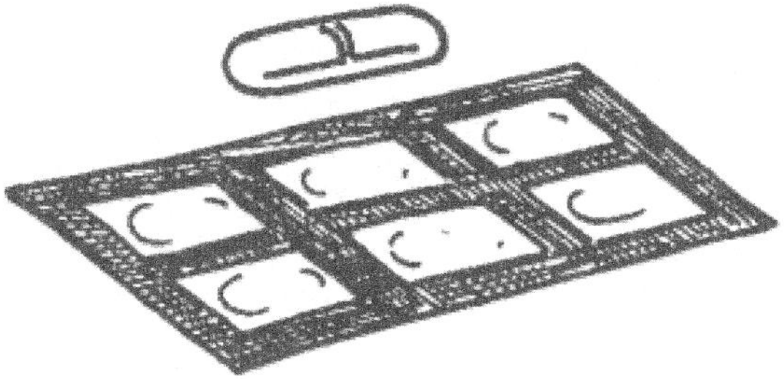

3. Tablets and capsules

Tablets are also called *pills*. Large tablets are called *bolus*. Capsules are small shells that have powdered medicine inside. They may be small, large, round, or elongated. Size and colour of tablets and capsules may also vary.

Examples: piperazine, Distodin®, Panacur®, sulfadimidine.

3. Ointments

Ointments are of a semi-solid consistency like soft butter. They are for external use only.

Examples: Himax®, Terramycin®

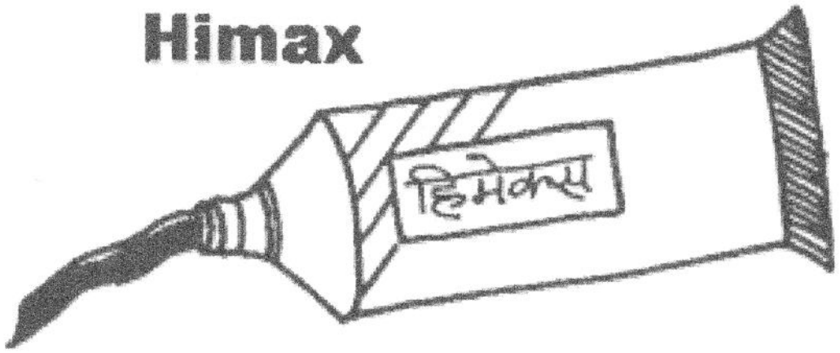

Injections:

Injections come in both powder and liquid forms. These drugs are injected under the skin or into the muscle using a syringe and needle.

Examples: Terramycin®, procaine penicillin

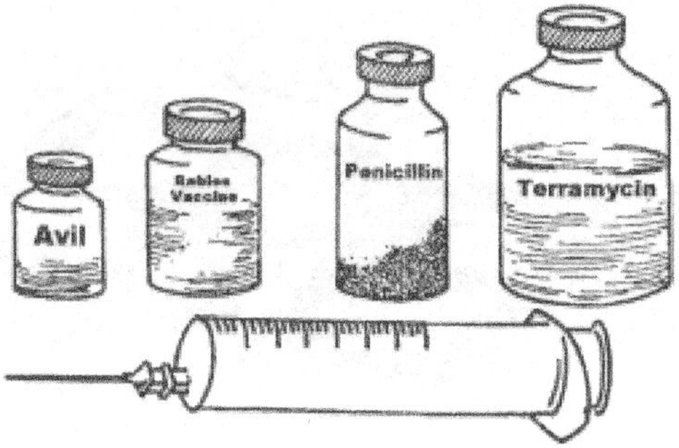

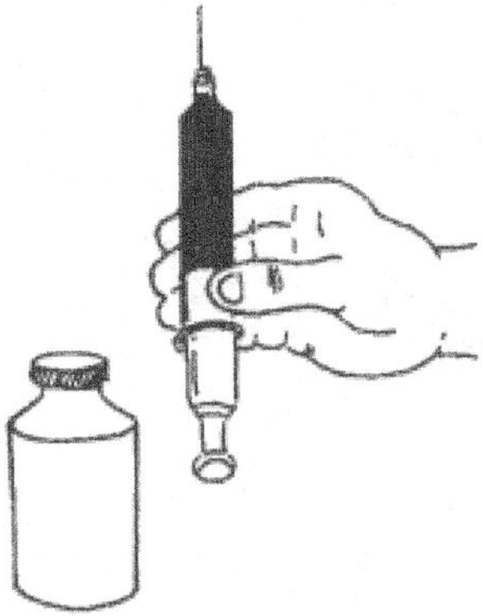

Part II: Dose of medication

- Dose means the amount of medicine used at one time. The dose depends on the weight of the animal and the kind of animal.
- Course of Medication means the amount medicine given daily for a number of days.

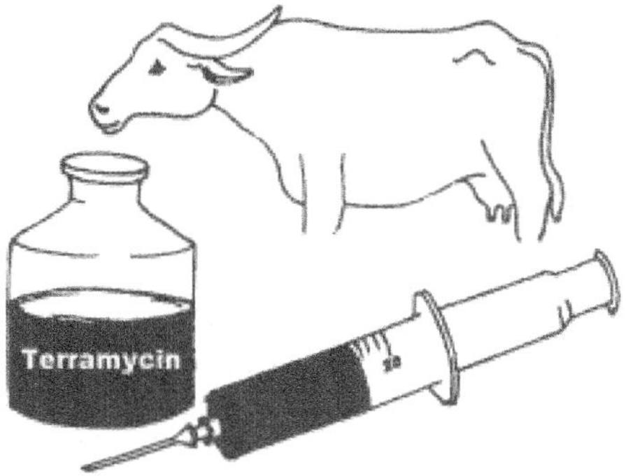

1. Doses based on weight

a) Large animal

Large animals need larger amounts of medicine

b) Small animals

Since the weight of this buffalo calf is less than the weight of an adult buffalo, less medicine is needed. Small animals need less medicine than big animals.

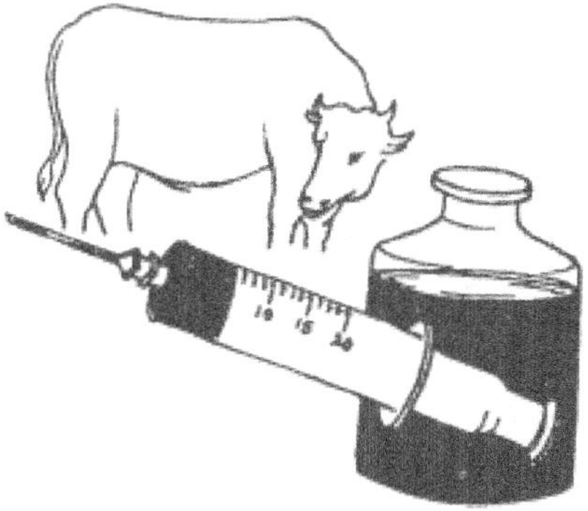

2. Doses based on the kind of animal

Even when body weight is the same, a calf and a dog need different doses of medicine.

For example: this calf weighs about 20 kilograms. The dose of piperazine is 20 tablets.

This dog also weighs about 20 kilograms, but for this dog, 10 tablets of piperazine is sufficient.

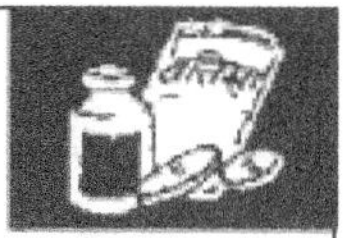

Why is the correct dose important?

- The correct dose will help the animal get well quickly.
- A dose that is too low will not kill the disease-causing organisms.
- A dose that is too high is expensive and might even kill the animal.

How do we measure the correct dose?

There are many ways to measure medicine. How to measure depends on the form of the medicine.

1. Powder medicine

It is necessary to measure powder medicine in grams (g) and milligrams (mg). Since balances are expensive, we can use a standard chewing tobacco can or plastic box to measure powder medicine. (See last page of Medicine book.)

Method of dividing a large quantity into small equal doses:

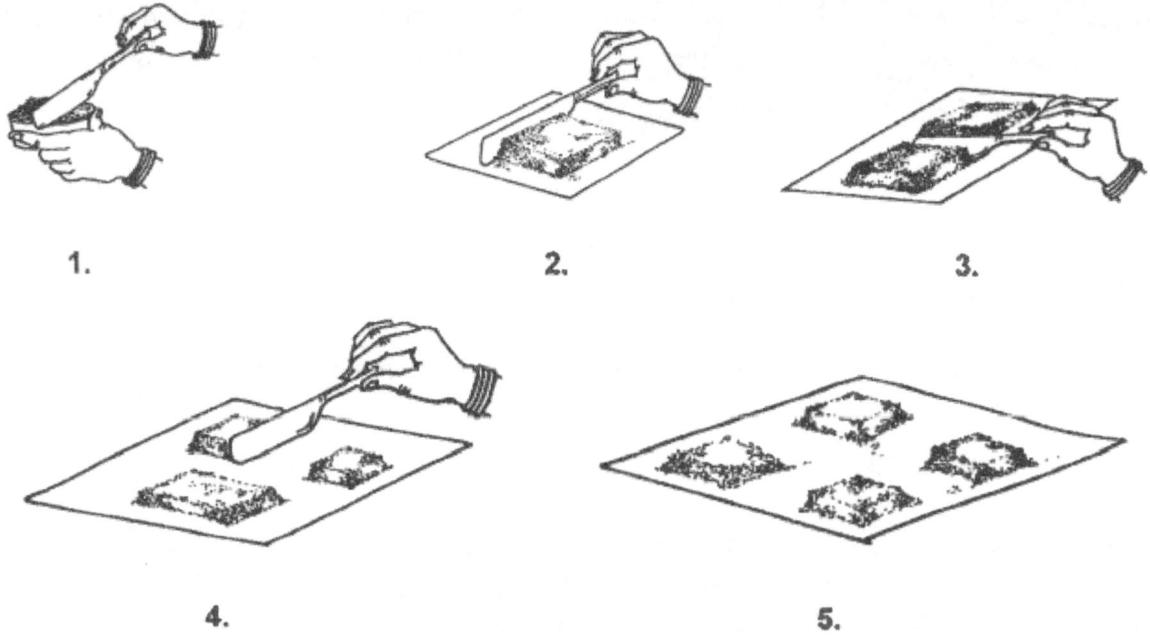

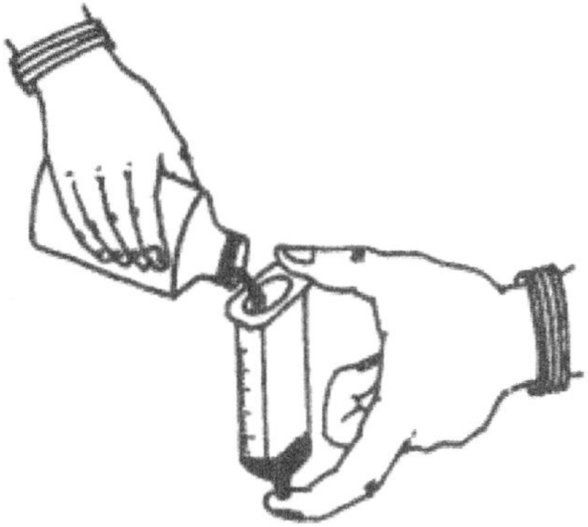

2. Liquid medicine

Liquid medicine can be measured in a syringe. Immediately after measuring the medicine, the syringe should be cleaned.

3. Tablets and capsules

It is easy to measure a dose with tablets or capsules. You can simply count out the required amount.

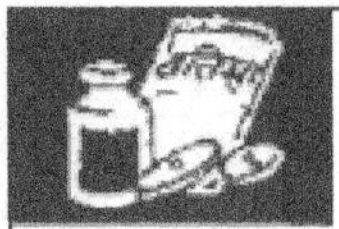

4. Ointments

Ointments should be applied in sufficient quantity to cover the wound.

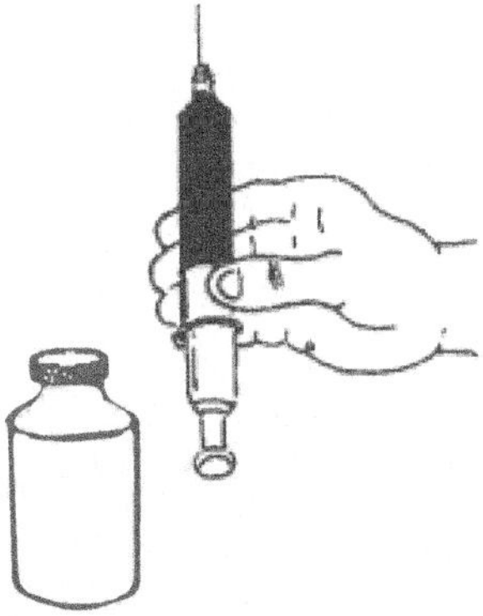

5. Injections

These can be measured directly in the syringe.

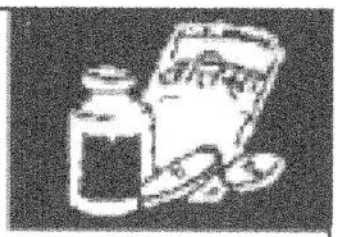

Part III: Preparation of Medicine

Many methods can be used according to the form of the medicine and the kind of animal.

1. Powder medicine:

After mixing the powder medicine with water (i.e. making it liquid medicine), it can be given with a bottle or a bamboo pipe. This method is called *drenching.* Powder medicines can also be mixed into special food. (For example: flour, corn husks, or porridge). For pigs and chickens, it can be mixed into their concentrated feed.

2. Liquid medicine

A drenching pipe (bamboo or plastic) is very useful. The liquid medicine is usually mixed with water and fed to the animal with the pipe. Pigs and chickens are difficult to drench. Their medicine is usually mixed in with the drinking water, care must be taken to make sure they drink the medicated water.

We must be very careful while drenching animals, especially small ones like goats and sheep, if the head is held too high, they cannot swallow. The liquid can enter the lungs and kill them. For this reason, we can use a rubber pipe, funnel, and bamboo or plastic pipe as shown. The rubber pipe goes directly into the stomach.

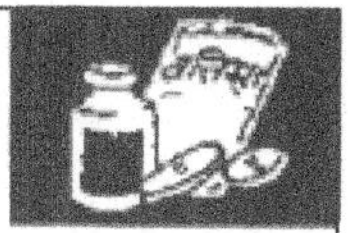

3. Tablet medicines:

- They can be wrapped in tasty leaves.
- They can be mixed with grain, flour, or molasses.
- Pigs and chickens can be fed by crushing the tablets and mixing them into the feed.

Making a medicine packet for the animal's owner:

During our spare time, we can prepare one-dose packets of various medicines. There are some advantages to having these medicine packets made up ahead of time:

- They are easy to carry when visiting sick animals.
- It makes it possible to deliver good, fast service even during busy times.
- Medicines can be kept safe from moisture, heat, mice, and children.
- Good use of free time.

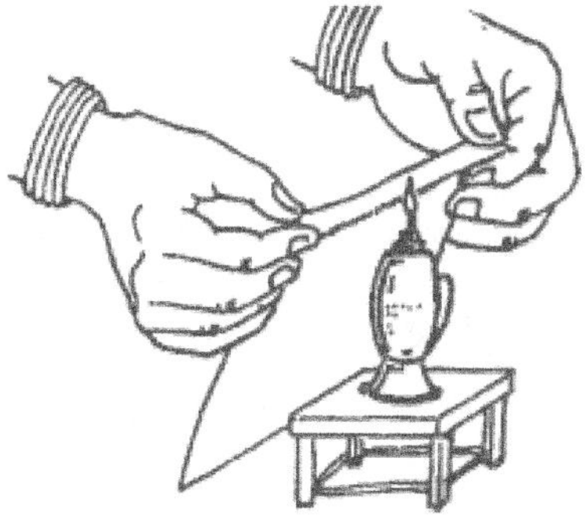

To make a packet of medicine, scrap paper or plastic can be used. Plastic is better than paper because it is waterproof.

There should be a clearly written label on each packet. The label should carry the following information:

1. Name of the medicine
2. Amount of the medicine
3. Expiration date (month and year)

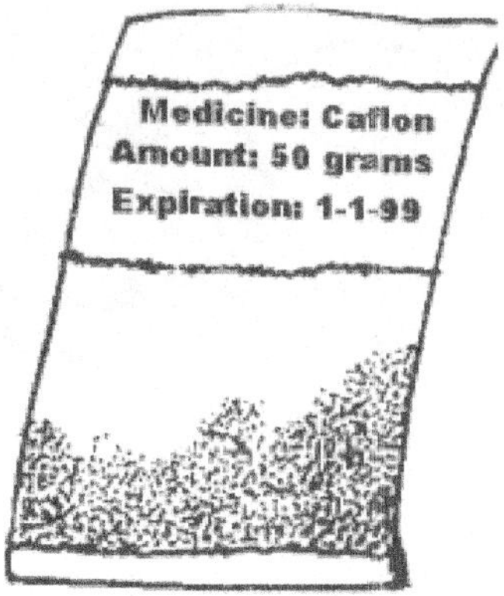

After making the medicine packets, place them in a big plastic bag and tie the top tightly. Always store the medicine in the medicine box. Remove expired medicines and dispose of them. Replace medicine with new stock as needed. Label all medicines clearly with their name and expiration date.

Part IV: Administration of Medicine

There are three basic methods of giving medicine.

1. By mouth:

After mixing powder, liquid, or tablet medicine with water, it can be given with a bamboo or plastic pipe. Large animals can be fed the medicine wrapped in leaves or mixed with flour. Sheep and goats should be fed liquid medicine by a rubber tube passed into the stomach.

How to do it?

- First clean all necessary items like the rubber tube, bamboo pipe, and funnel.
- Measure the distance from the nose to the last rib and mark on the rubber tube.
- Place the bamboo pipe in the goat's mouth and hold the mouth closed. The bamboo pipe prevents the goat from chewing the rubber tube.
- Insert the rubber tube gently up to the mark.
- To make sure the tube is in the stomach, have someone listen while you blow into the tube. There will also be a bad smell from the tube if it is in the stomach.
- Attach the funnel to the tube and pour the medicine in slowly.
- When finished, slowly remove' the tube first, and then remove the bamboo pipe.

Precautions with giving medicine by mouth:

The animal will cough if the rubber tube or liquid goes into the air pipe. If you think the tube is in the air pipe, remove it, and try again in a few minutes after the animal has had a chance to rest.

2. By External Application:

For skin, eye, and ear problems, ointments are used. They should be applied to completely cover the affected area. e.g. Himax, Charmil

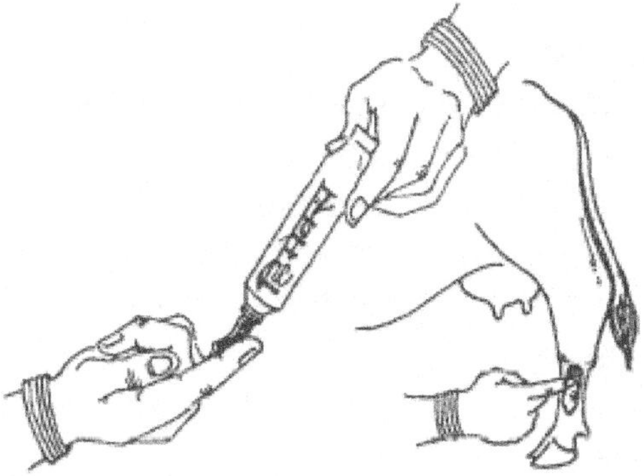

3. By Injections:

Injections are given under the skin or in the muscle e.g. Teramycin, Avil. Powder medicines for injection must be mixed with special injection water before using e.g. Procaine, Penicillin, Penidure.

Precautions for handling poisonous medicines

Some medicines, especially ones used to kill external parasites, can also be dangerous to us! If these medicines are not handled properly they can cause sickness and death in both people and animals. Examples: malathion, Butox®, and Neocidol®.

The following precautions should be followed while handling poisonous medications:

- The person applying the medication should cover his nose and mouth with a cloth.
- Avoid getting the medication into the animal's eyes, nose, or mouth.
- A muzzle can be used to prevent the animal from licking medication from its skin.
- Do not apply powdered medicine (dust) when it is windy.

- After using poisonous medicines, wash yourself thoroughly.
- Move the animal away from the shed and food sources before applying poisonous medication.
- Do not store poisonous medicines with food.
- Do not re-use the bags or bottles of poisonous medicines. They should be buried.
- Expired, unused poisonous medicine should be buried. Do not dispose of it near water sources.
- Keep poisonous medicine out of reach of children.

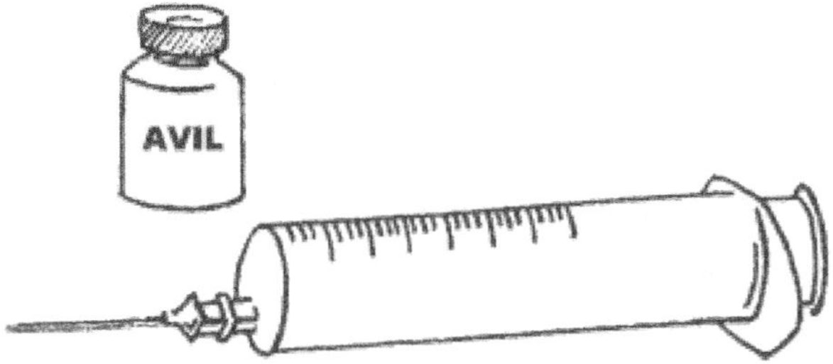

General Precautions

- Keep all medicines out of the reach of children.
- If an animal has become sick from the medicine, treat it as you would for poisonous plant ingestion.
- If a person has become sick from the medicine, contact the local health worker immediately.

Injections

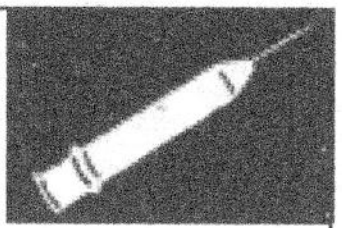

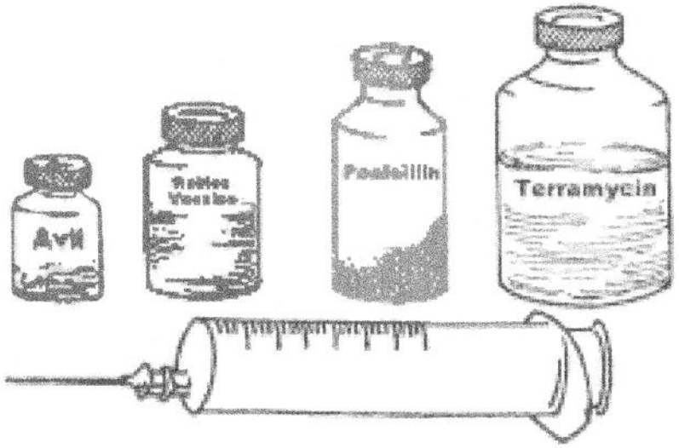

Injections are used for both prevention and treatment. Injectable medicines work quickly. Many medicines can be given by injection, such as: vitamins, anti-fever medicine (antibiotics), and vaccines.

Preparation of injections:

1. Cleaning and sterilizing the needle and syringe

- Needle and syringe should be cleaned either with Dettol® or Savlon solution. If these are not available, local alcohol or boiling water can be used.
- After cleaning with Dettol® or Savlon®, rinse with boiled water.
- Keep needles and syringes in a clean place.
- Dettol® or Savlon® solution can be kept and used for 1 week.
- If our equipment is not clean, the injection site may become infected!

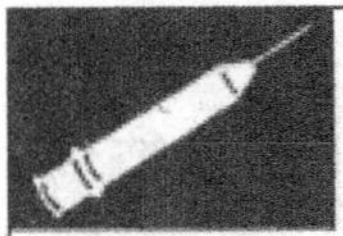

2. Make a dose of injectable medicine:

- We can measure the dose of liquid medicine directly into the syringe.
- Powder medicine doses are measured only after mixing with water.

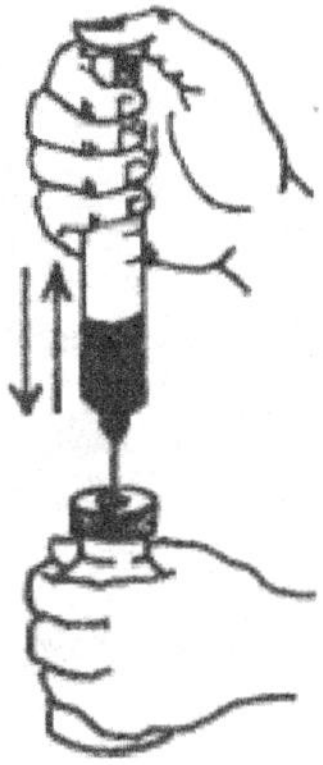

- Insert the syringe into the vial as shown.

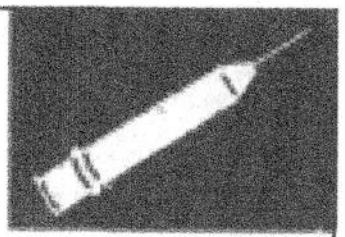

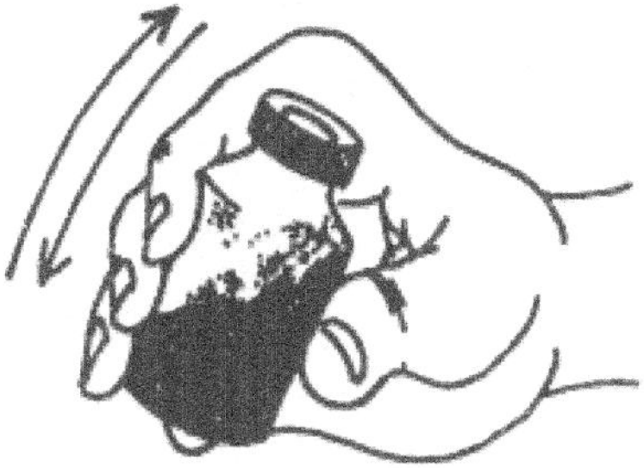

- Slowly take air into the syringe as the water is put into the vial.
- Pull the needle out of the vial.
- The medicine must be mixed well! Shake the vial gently to mix the water and the medicine. Make sure your hand does not touch the top of the vial.

3. How to get the medicine into the syringe?

- Fill the syringe with the same number of cc's of air as the number of cc's of medicine you want to take out.
- Push the needle into the vial.
- Hold the syringe below the vial as shown.
- Slowly add some air and withdraw a little bit of medicine. Repeat this process until the entire dose is in the syringe.

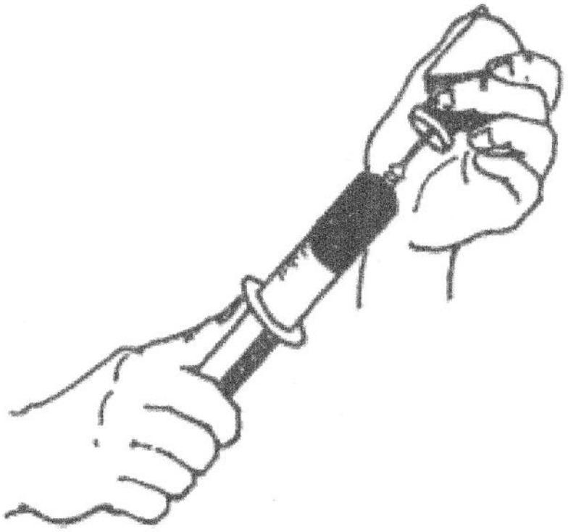

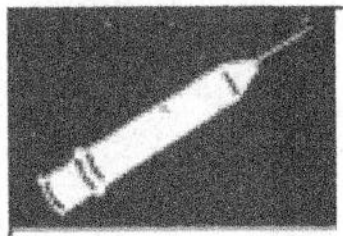

4. How do we remove air from the syringe?

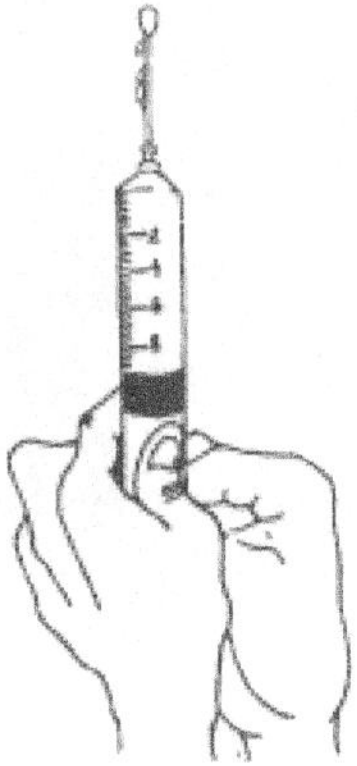

- Hold the syringe upright (needle pointing up).
- Tap the syringe gently to move any air bubbles to the top.
- Remove all the air from the syringe by gently pushing up on the plunger.

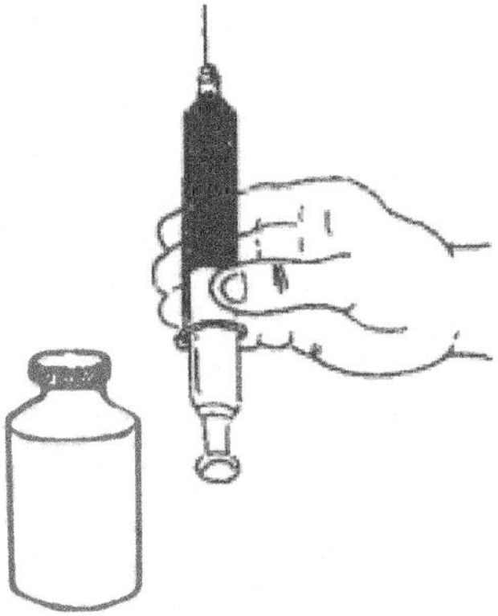

5. Dividing larger dose into a smaller dose

Sometimes a large quantity of medicine must be divided into a smaller dose.

For example, Sometimes we need only 100,000 units of penicillin but the vial contains 400,000 units. Similarly, by dividing the dose, we can make a dose of 600,000 units from vial that contains 2,400,000 units.

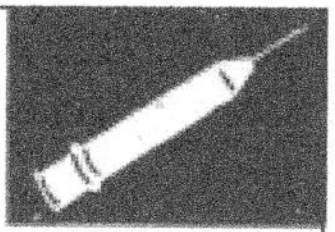

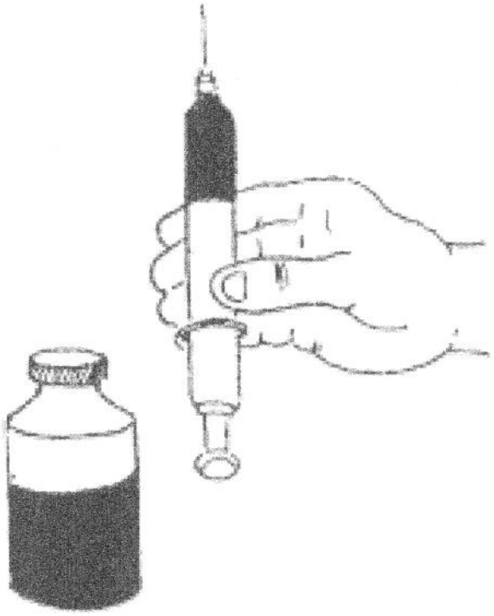

For example: It is easy to divide 10 ml volume of 400,000 unit penicillin.

If we mix about 8 ml of water with the powdered penicillin, that will give us about 10 ml of liquid medicine. If we then divide this 10 ml into 2 parts so that 5 ml is in the syringe and 5 ml remains in the vial, we will have a dose of 200,000 units in the syringe.

Again, if we divide the 5 ml in the syringe into 2 parts, and put 2.5 ml back into the vial, then the syringe will contain 100,000 units of penicillin.

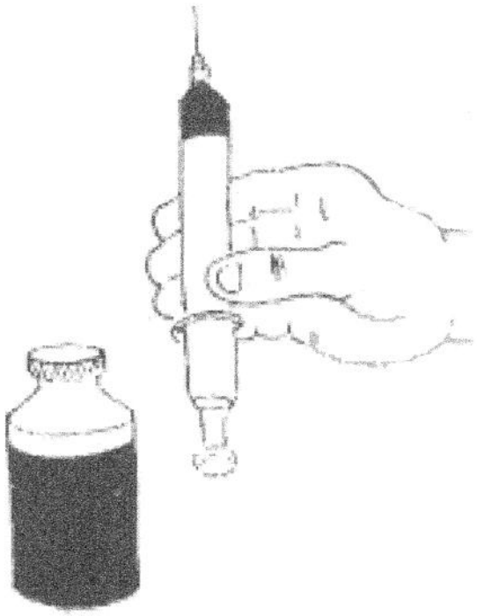

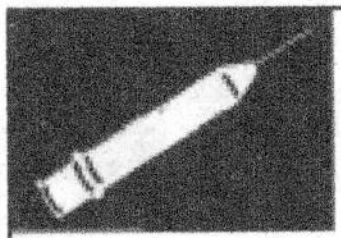

Methods of giving injections:

Injections should be given when the animal is fully restrained.

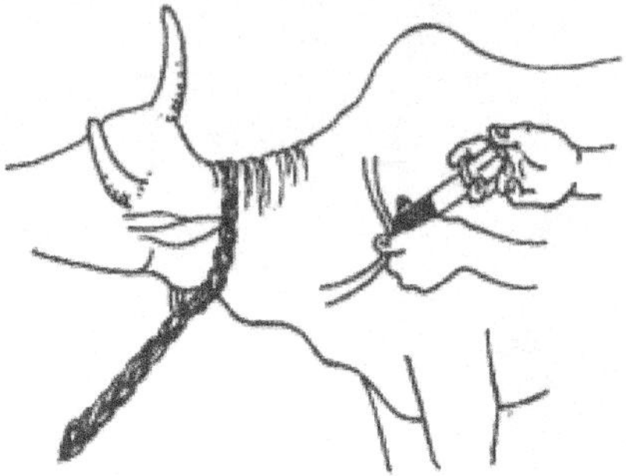

There are 3 main methods of giving an injection:

1. Under the skin *(subcutaneous* or *SC)*
2. Into the muscle *(intramuscular or IM)*
3. In the blood vessel *(Intravenous* or *IV)*

- Vaccines are usually given under the skin

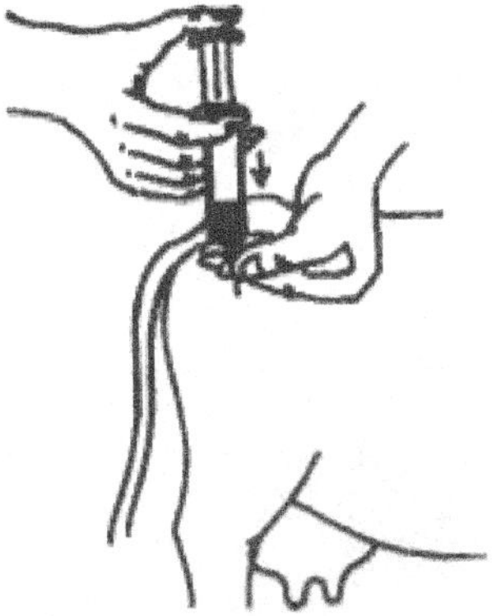

- Vitamins and medicine for fever *(antibiotics)* are given into the muscle

- Only a veterinary doctor should give injections into the blood vessel *(intravenous)).*

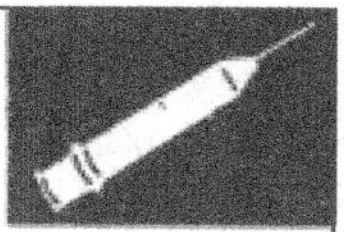

Where to inject?
The injection site varies according to the kind of animal

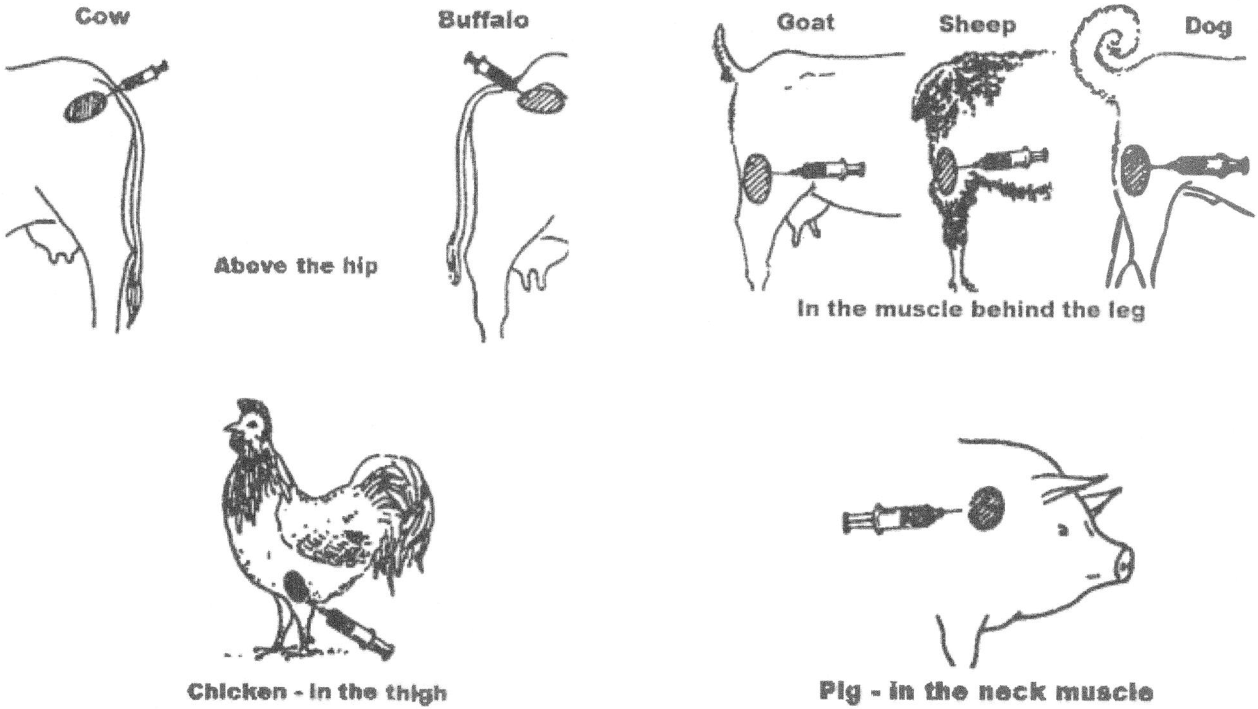

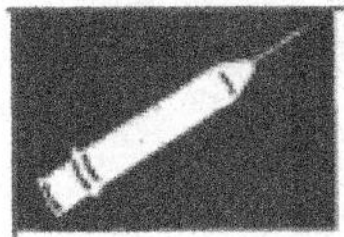

How do we inject?
Intramuscular injection technique:

1. Choose a site for injection as shown in the picture.

2. Clean the skin using Dettol®, Savlon®, or alcohol.

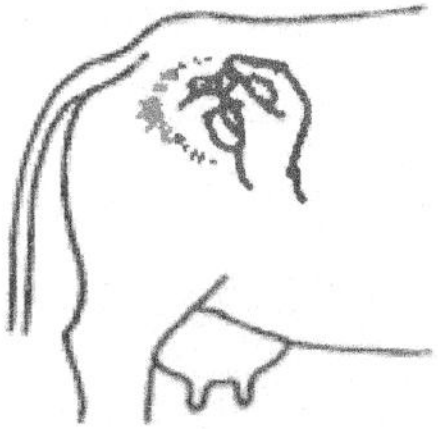

3. Hit the animal 3 times with the hand then insert the needle as far as it will go.

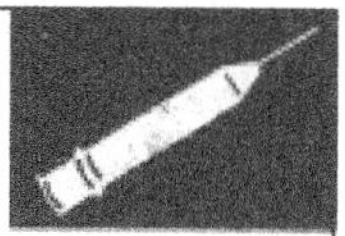

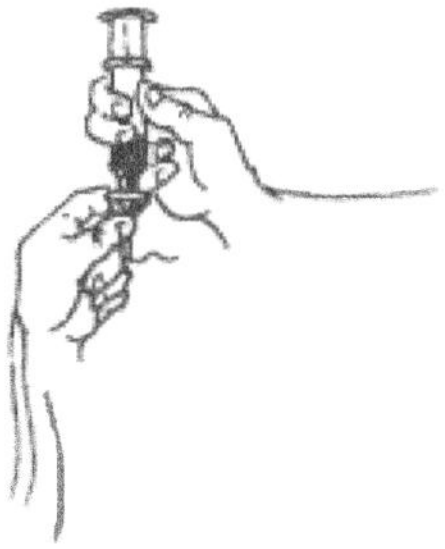

4. Attach the syringe tightly to the needle.

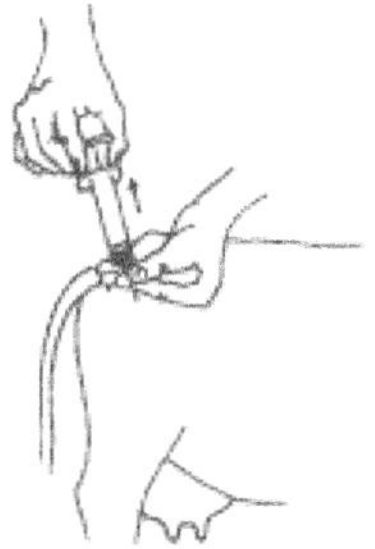

5. Check to see that the needle is not in a blood vessel. To do this, pull the syringe plunger a little bit. If no blood comes into the syringe, gently push the plunger to inject the dose. If you see blood in the syringe, remove the syringe and needle and choose a new site.

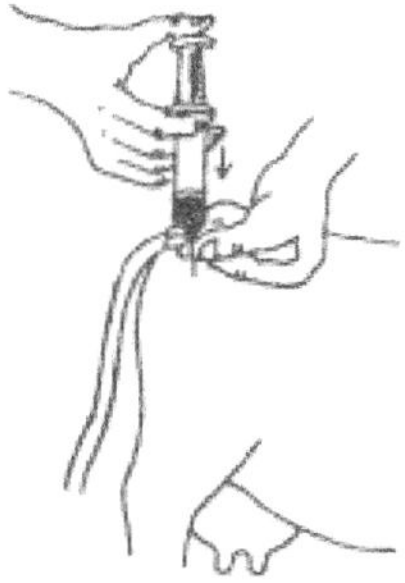

6. After removing the needle, gently massage the injection site.

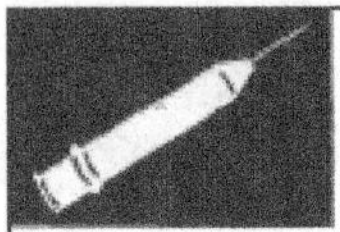

Techniques for injecting under the skin

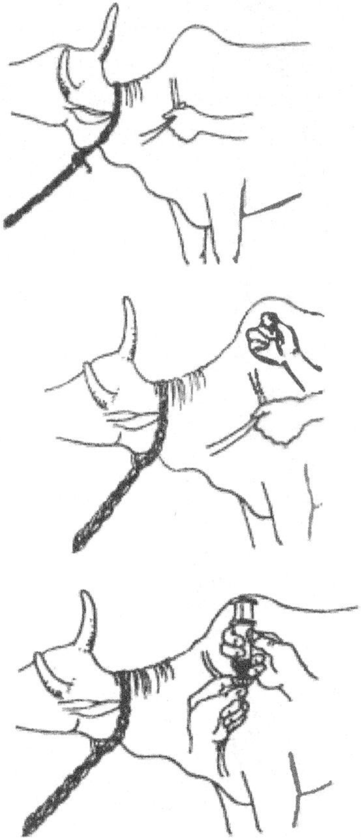

There are 2 techniques:
a) Pulling the skin then inserting the needle.
b) Inserting the needle first, then pulling the skin.

1. Subcutaneous injections are given under the loose skin of the neck area near the point of the shoulder.

2. We must clean the injection site before placing the needle. Pull up on the loose skin and insert the needle. Or, you can insert the needle as you would for an intramuscular injection, then pull outward on the skin and needle together. In cases where a live vaccine is being used do not clean the site of injection by antiseptic.

3. Attach the syringe tightly to the needle and pull outward on the skin and needle together.

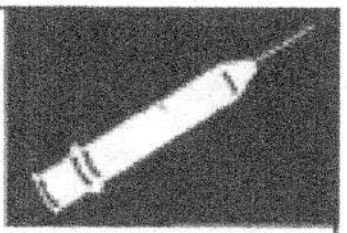

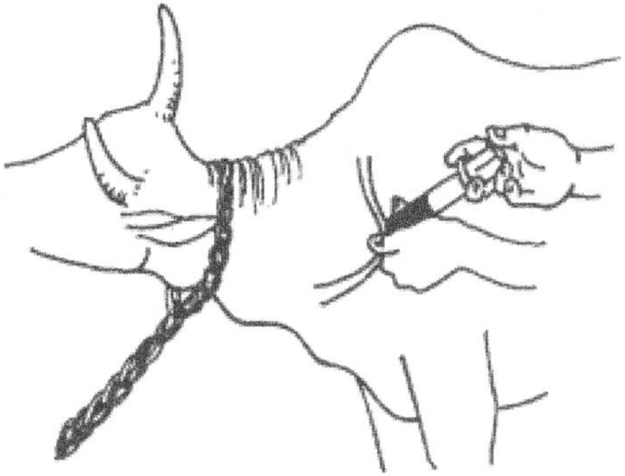

Inject the medicine under the skin. It should go in easily. If it does not, the needle is not in the right place. After injecting the medicine, remove the needle and massage the injection site.

When you are finished, clean the syringe and needle and store it in a safe place. Tap water can be used to clean the syringe.

Injection needles:

There are various sizes of injection needles available. Some are thick, some are thin, some short and some long. Needles are chosen according to the size of the animal and the kind of medicine being used.

For example:

- An 18 gauge 1½ inch long needle is used in cattle and buffaloes.
- A 20 gauge 1 or ½ inch long needle is used for sheep goats, and pigs.
- A 21 gauge 1 inch long needle is good for use on chickens.
- A 16 or 18 gauge 1 inch long needle is good for subcutaneous injections in cows and buffaloes.

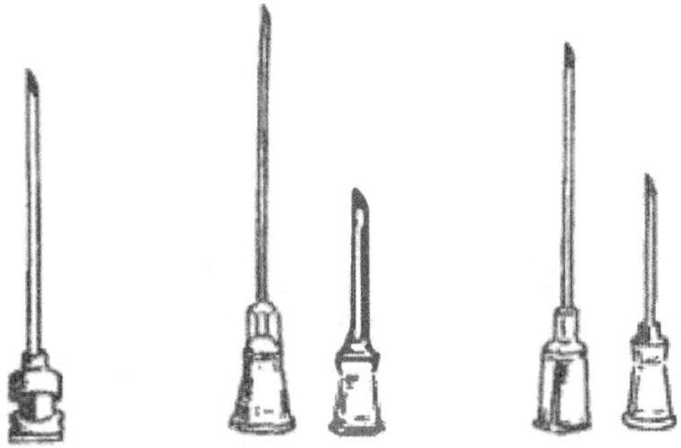

16 ga **18 ga** **21 ga**

Antibiotics

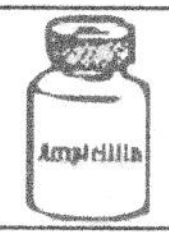

What is an Antibiotic?

An antibiotic is the product of microorganisms which act against bacteria.

How do they work?

Bacteriostatic Antibiotics

These antibiotics stop the growth of bacteria.

Examples are:

- Sulphonamides
- Tetracycline
- Chloramphenicol
- Erythromycin

Bactericidal Antibiotics

These antibiotics kill the bacteria.

Examples are:

- Penicillins
- Streptomycin
- Neomycin
- Cephalosporins

> **NOTE:**
> NEVER GIVE A COMBINATION OF BACTERIOSTATIC AND BACTERICIDAL ANTIBIOTICS. THERE WILL BE NO OR ADVERSE EFFECT OF ANTIBIOTICS.

Illustration

Penicillin (Bactericidal)	+	Tetracycline (Bacteriostatic)	=	**ANTAGONISTIC**
Penicillin (Bactericidal)	+	Streptomycin (Bactericidal)	=	**SYNERGISTIC**

ANTIBIOTIC SPECTRUM

It is the field of antibacterial activity of an antibiotic.

1. Narrow Spectrum Antibiotics

Antibacterial effect is restricted to a relatively small number of organisms.

For example:

Penicillin is mainly active against Gram Positive bacteria.

2. Broad Spectrum Antibiotic

Antibacterial effect is against a wide variety of organisms, such as Gram positive and Gram negative bacteria.

For example: Chloramphenicol, Oxytetracycline, Ampicillin, Penicillin + Streptomycin

An example of three antibiotics are:

1. Strepto-Penicillin

When do we use it ?

- Fever, retained placenta
- Uterine infection
- Mastitis
- Dystocia, pneumonia
- Deep infected or maggoted wounds
- Actinobacillosis and actinomycosis
- Foot rot

Dose

- Add 7.5 ml. of distilled water to make 10 ml. suspension.
- Cattle, Buffaloes and Horses - 1 ml. per 25 kg. body weight.
- Sheep, Goats, Pigs, Dogs - 1 ml. per 5 kg. body weight.

How many times do we give the dose?

One dose daily for 3-5 days.

How do we give it?

Inject into the muscle.

What forms of medicines are available?

Powder form in small vials. Each vial contains 2.5 grams of medicine.

2. Ampicillin

When do we use it?

- Pneumonia, respiratory tract infection
- Urinary tract infection
- Uterine infection
- Gastro-intestinal infection (gastritis and enteritis)
- Ear infection
- Mastitis
- Calf Scour
- Swine erysipelas in pigs
- Foot abscess

Dose

- Cattle, Buffalo, Sheep, Goat, Pig, Horse - 10 mg. per kg. body weight orally and 7 mg. per kg. body weight intramuscularly, intravenously or subcutaneously.

Precaution

- The medicine solution in distilled water should be used within 3-4 hours of preparation.

How many times do we give the dose?

- One dose twice daily.

How do we give it?

- Through the mouth (orally), into the muscles (intramuscularly), under the skin (subcutaneously) or into the veins (intravenously).

What forms of medicines are available?

- They are available in tablets, powder, or solution form. The amount varies according to the forms available.

3. Amoxycillin

When do we use it?

- Pneumonia, respiratory tract infection
- Urinary tract infection
- Uterine infection
- Gastro-intestinal infection (gastritis and enteritis)
- Ear infection
- Mastitis
- Calf scour
- Swine erysipelas in pigs
- Foot abscess

Dose

- 10 mg. per kg. body weight.

How many times do we give the dose?

- One dose twice daily.

How do we give it?

- Through the mouth (orally), into the muscles (intramuscularly), or into the veins (intravenously).

What forms of medicines are available?

- They are available in tablets, powder, or solution form. The amount varies according to the forms available.

Why medicine does not always work?

A farmer's animal was examined and treated by an experienced Village Animal Health Worker. Many types of expensive medicine were given, but the animal's condition did not improve.

Why did the medicine not work?

There are many possible reasons for the failure of medical treatment.

1. The animal was brought too late for treatment.

This is not the fault of the Village Animal Health Worker but it is the fault of the owner. Treatment may not work once the condition has become very severe or long-standing.

2. Insufficient dose of medicine

The dose is made according to the body weight of the animal. If the estimated dose is incorrect (too small), the treatment may not work.

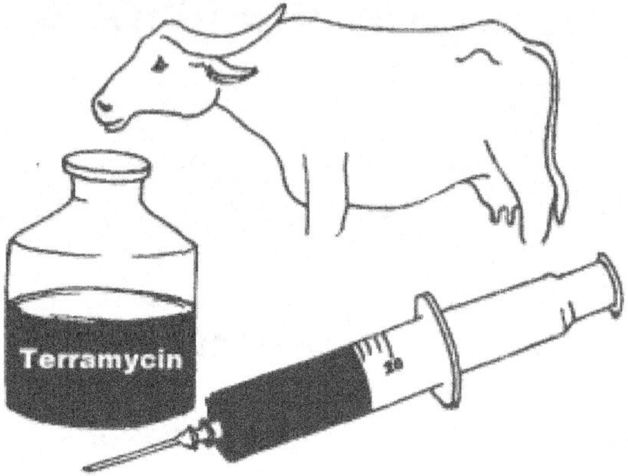

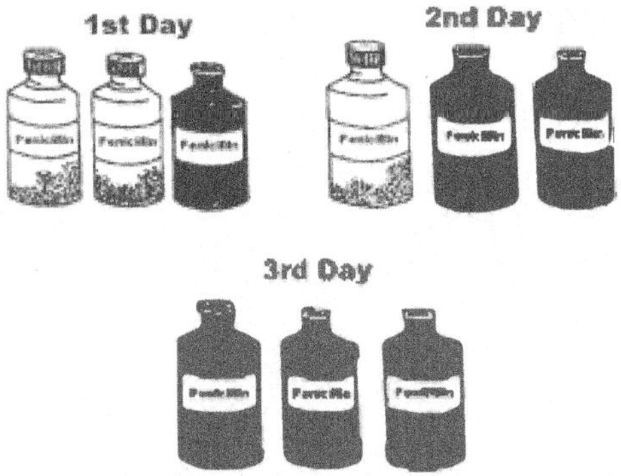

3. Medicine was not given regularly

The required dose must be given at specific, regular times. This must be explained to the owner before starting the treatment. If the medicine is not given regularly, or if the medicine is stopped too soon, the treatment may not work.

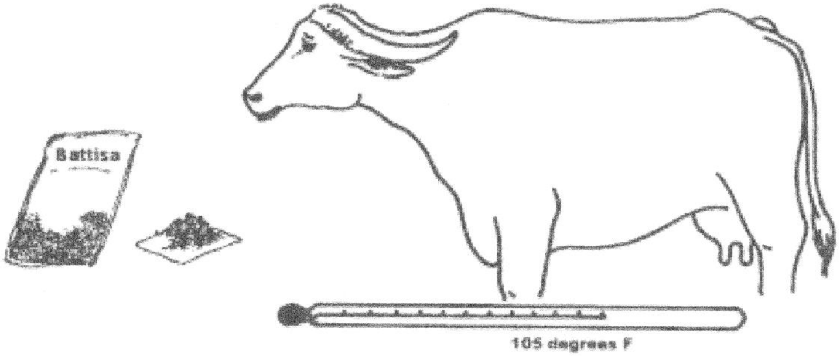

4. The correct medicine was not used.

Different types of medicine should be used depending on the disease and the kind of animal. If the correct medicine for the particular disease is not used, the animal may not get better.

5. Damaged medicine

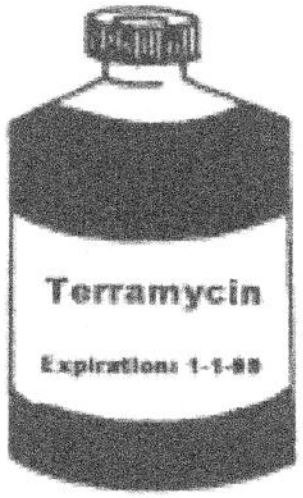

Medications have a specific “shelf life” after which time they do not work well. Most manufacturers stamp their medicine with an *expiration date.* After this date the medication should not be used.

Some medicine may become damaged due to carelessness. They may become damaged before the expiration date if they have been exposed to lots of heat, sunlight, or water. Damaged medicine does not work and can be dangerous.

6. Some diseases have no treatment:

Not all diseases are treatable. For example, there is no treatment for rabies once clinical signs appear. The animal will die. Some diseases like emphysema or heaves usually do not cause death in affected animals, but medicine will only help a little bit. It will not cure them.

7. Lack of good supportive care

Medicine alone will not work well without good supportive care. Without good supportive care, the animal might even get worse instead of better, even though we gave them lots of medicine.

Nursing Care for the Sick Animal

What is nursing (supportive) care?

When you are ill, sometimes it is necessary to be hospitalized. What advice do the doctor and nurse give to their patients?

The advice would be similar to the following:

- Eat fish, meat, milk, and eggs.
- Eat lots of fruits and green vegetables.
- Get plenty of rest.
- Sleep in a clean bed.
- Wear clean clothes.
- Use medicines at the appropriate times.

Similarly, appropriate supportive care must be provided for our sick animals if they are to get better.

- Feed high quality green feeds
- Provide sufficient water.
- Give the animals rest.
- Keep the animals in a well-ventilated shed.
- Keep the shed dry.
- Feed grains and concentrates.
- Keep the sick animal away from the healthy ones.
- Give medicine at the appropriate times.

Without good supportive care, medicine alone will not work well. It will be more costly if we depend only on medicine and ignore good supportive care.

Fever

Livestock, birds, and humans have warm bodies. This warmth is called their *normal body temperature*. If the temperature is higher than normal because of disease, then this is called *fever.*

Normal body temperatures of livestock, birds, and humans:

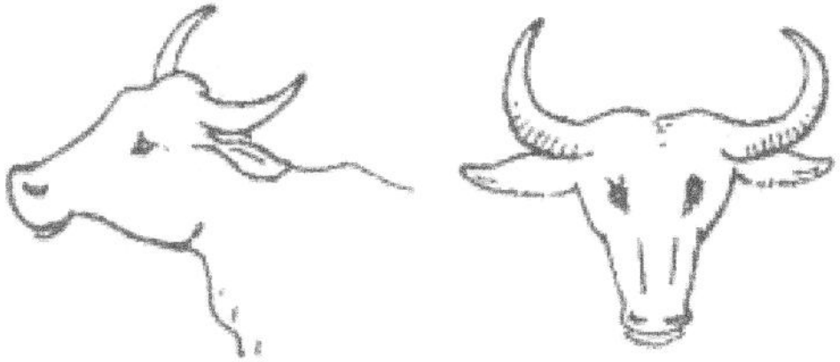

102°

103°

100°

106°

99°

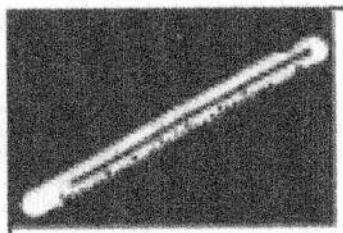

Walking, working, and grazing in the sun can cause the body temperature to rise. This type of increase in temperature is not a fever.

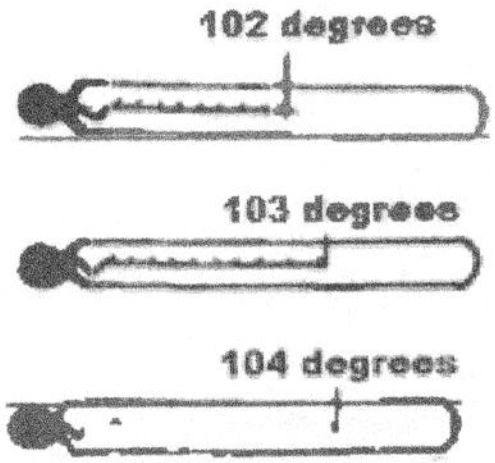

This animal's temperature could be...

102°

103°

more than 104°

Note: If an animal's temperature has only risen between 1 and 3 degrees, and the animal is showing signs of an illness, this increase is probably due to a chronic infection. These animals generally need a long course of antibiotics to clear the infection.

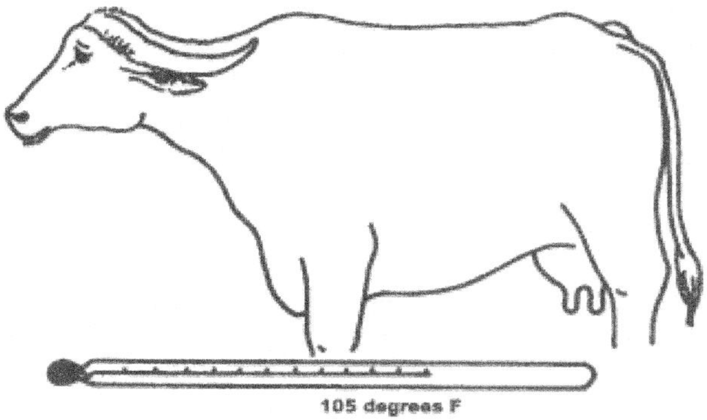

When the temperature of the body rises due to bacteria or viruses attacking it, this is called a fever.

Signs of fever:

- Not eating
- Dry muzzle
- Rapid breathing
- Dehydration
- Decreased milk production
- Yellow colour of the urine
- Shivering even when the weather is warm

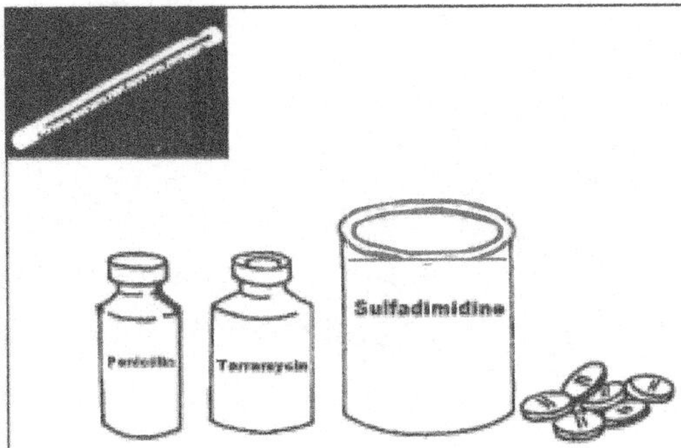

Treatment of fever:

- Procaine penicillin or Penidur®
- Terramycin®
- Sulfadimidine
- Feverdon® powder
- Herbina® capsules

Note: If an animal's temperature has increased by more than 3 degrees this generally is considered a high temperature, and due to an acute infection. Therefore anti-inflammatory (anti-fever) medication as well as a strong course of antibiotics may need to be used.

Cover the body with a wet cloth to give the animal some relief.

Special precautions while treating:

Medication for treatment of a fever must be given for 3 to 5 days in a row. If the treatment is cut short, it may not work. The fever may return and it will be harder to treat the second time. This adds to the expense of treatment and the animal might die.

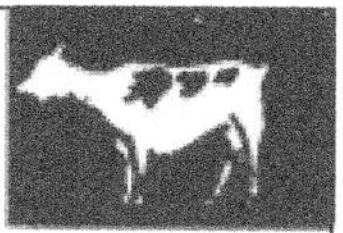

Wounds

Many different kinds of wounds can affect the body.

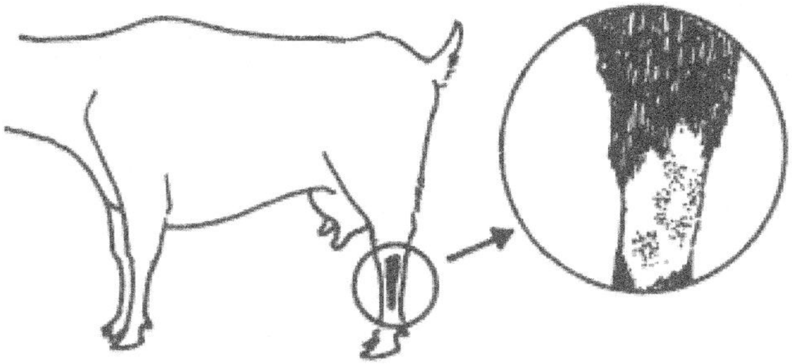

Causes of wounds:

- Falling on sharp rocks
- Kicking or biting
- Burns from fires
- Scratches
- Broken horns from fighting
- Cuts and punctures from rocks, glass, or metal

Normally there are 3 types of wounds:

1. Fresh wound

When the skin and flesh have been cut within 1-2 days, this is called a *fresh wound.* This wound does not have a bad smell and is not maggot-infested.

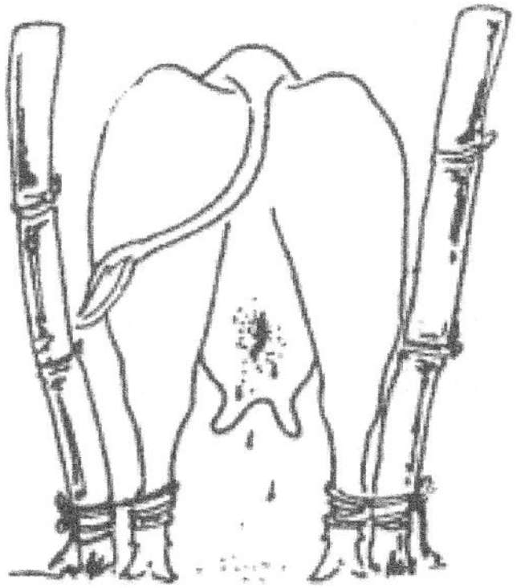

2. Infected wound

After a few days, a wound can become infected with bacteria. Pus will develop. This wound smells bad. This type of wound is called an *infected wound.*

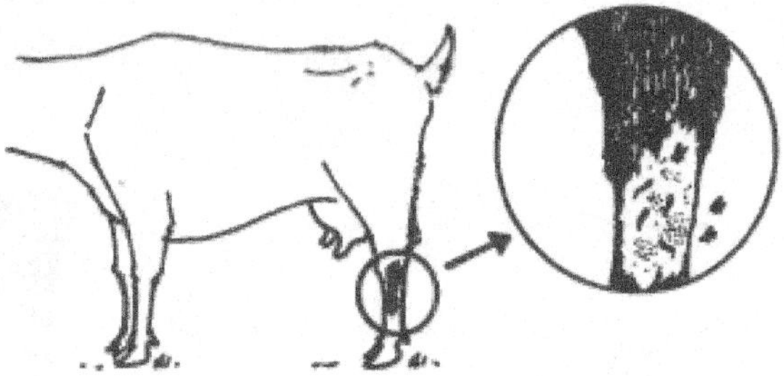

3. Maggot-infested wounds

If a wound is not treated promptly, flies may lay eggs in the wound. From these eggs *maggots* (worms) develop. The maggots make the wound deeper and cause blood and body fluid to flow from this kind of wound. These maggot-infested wounds smell bad because they are also infected.

Treatment of wounds:

1. Treatment of fresh wounds

- If necessary, trim the hair away from the wound.
- Remove any foreign material from the wound.
- Clean the wound with Dettol® or Savlon® solution.
- Dry the wound with a clean cloth.
- Honey may also be used on wounds as an alternative to Himax ointment. Honey promotes skin growth and due to its high concentration of sugar it prevents bacteria from growing.

- Apply Himax® ointment to the wound.
- If possible, apply a bandage.
- If the wound is deep, treat as for a fever, with antibiotics.
- Continue to clean the wound and apply ointment daily until the wound is healed.

2. Treatment of infected wounds

- Trim the hair away from the wound if necessary.
- Clean the wound with Dettol® or Savlon® solution.
- Remove the pus.
- Flushing the wound with water from a syringe will help remove the pus.
- Remove all dead tissue with a clean forceps.
- Dry the wound and apply Himax® ointment.
- If a fever is present, or if the wound is very large or very deep, give penicillin injections.
- Continue to clean the wound and apply ointment each day until the wound is healed.

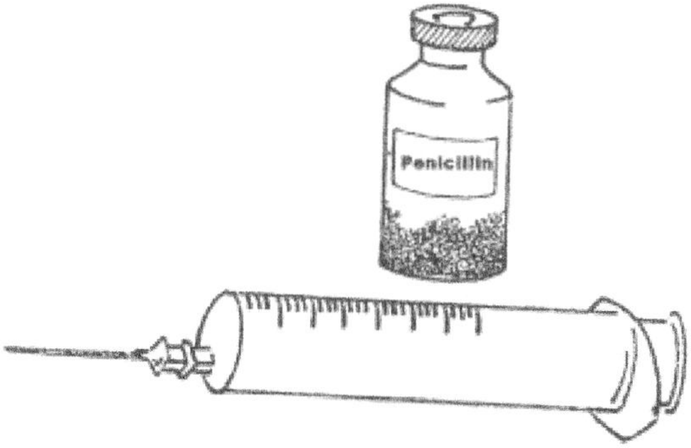

2. Treatment for maggot-infested wounds

- Kill the maggots using turpentine oil, naphthalene balls, CTC, or juice from new leaves of the peach tree.
- Within 5 to 10 minutes of applying the above medications, the maggots should drop off.
- If any maggots remain, remove them manually with the forceps.
- Clean the wound thoroughly with Dettol® or Savlon® solution.
- Dry the wound with a clean cloth.

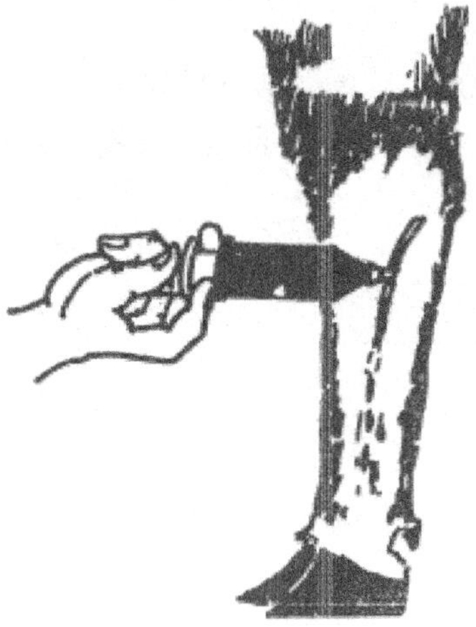

- Apply Himax® ointment to the wound and to the skin around it. This will help heal the wound and also prevent flies from laying more eggs on the skin.
- If the wound is large or deep, or if a fever is present, give penicillin injections.
- Continue to clean and treat the wound each day until it is healed.

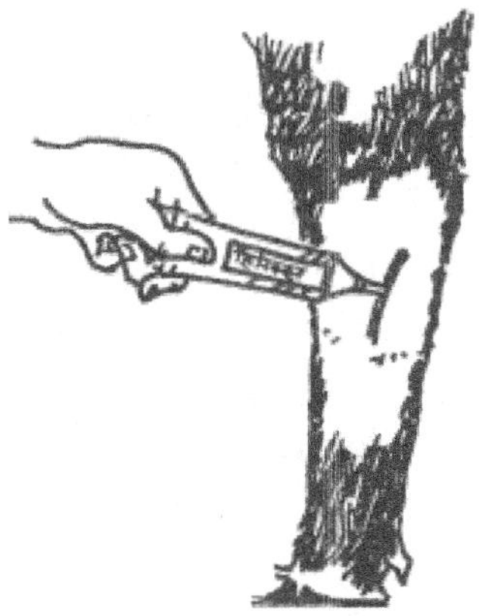

Bandaging

Why do we use a bandage?

- To keep the wound clean so it will not get infected
- To stop bleeding from a wound
- To decrease the pain caused by a fracture or sprain
- To prevent movement of a fractured bone

Materials needed to apply a bandage:

- Soft cloths or bandage material
- Antiseptic solution (Dettol® or Savlon®) or antiseptic powder, cotton
- Scissors

How to apply a bandage:

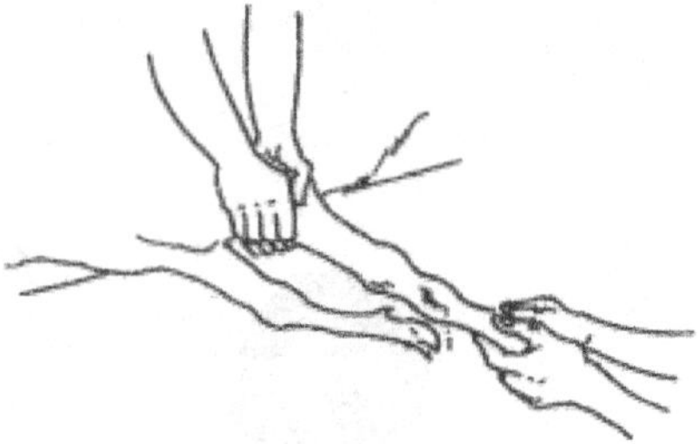

- If there is a wound, clean it with the antiseptic solution.
- Dry the wound with cotton or a clean cloth.
- Apply antiseptic ointment or powder to the wound.

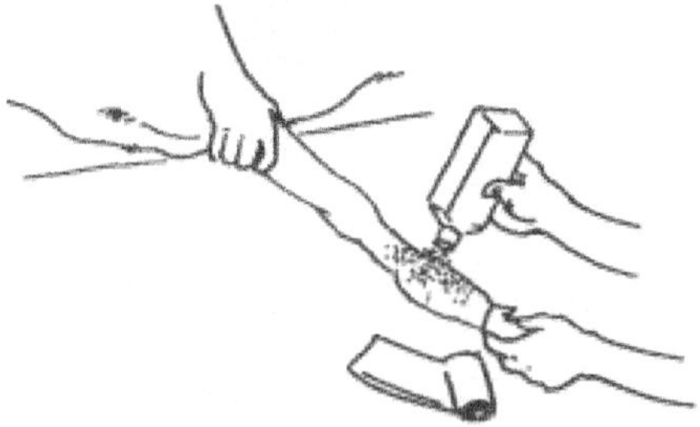

- To bandage a leg or a foot, where possible start from the joint below the wound and roll the bandage around the leg, covering about half the bandage cloth with each turn.
- Continue wrapping the bandage up the leg and stop at the joint above the wound (this helps to prevent the bandage slipping).
- Split the last 15 cm of bandage cloth down the middle to make 2 tying strips.

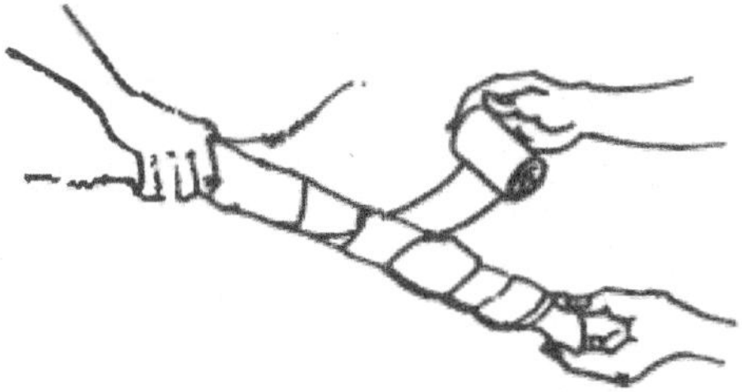

- Wrap one strip in the direction of the bandage wrapping and the other in the opposite direction.
- Tie the strips together snugly but not too tight. If tape is available, you can use it to help secure the bandage without tying.
- Check the bandage daily and change it if it gets wet, smelly, or too tight.

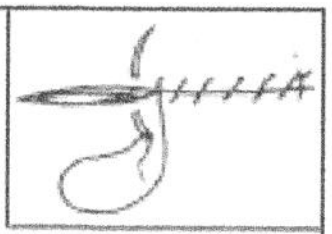

Suture Materials and Techniques

What is a Suture?

A suture can be defined as the technique used to close a wound.

Use of Sutures

- To close the edges of a wound
- To keep the medication in a wound for a long time
- To close a puncture/stab wound made by accident
- To prevent bleeding from a wound

Suture Materials

Suture materials are the threads used to close the edges of tissues. The suture materials can be classified into absorbable and non-absorbable depending on their ability to be absorbed in the tissues during the healing process.

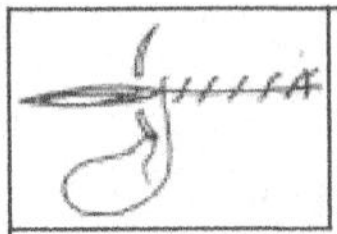

Absorbable Suture Materials

The absorbable suture materials are those suture materials which are organic in nature and are absorbed into the body tissues after a variable period of time. Absorbable suture materials are mostly used to close the internal organ/wound, e.g. catgut, fascia lata, kangaroo tendon, polyglactin 910 etc. Catgut is made from the sub mucosa of the small intestine of sheep and is the most widely used absorbable suture material in surgery.

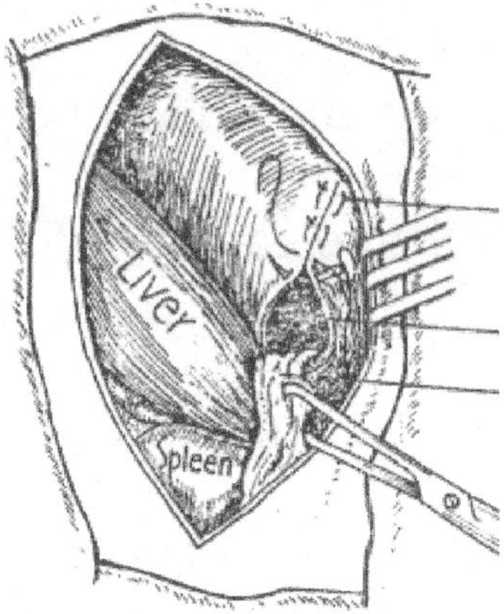

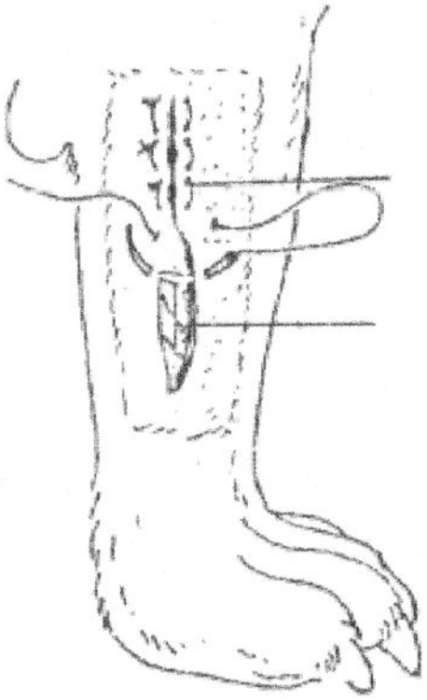

Non-absorbable Suture Materials

The non-absorbable suture materials are those suture materials which are not absorbed into the body tissues during healing of the wound. These suture materials are used to close the external wound (skin). e.g. cotton, silk, stainless steel wire, nylon thread etc. These sutures should be removed from the body after completion of the healing process, usually after one or two weeks.

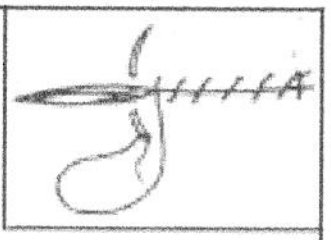

Suturing Techniques

Broadly, the suturing techniques can be categorized into two types:

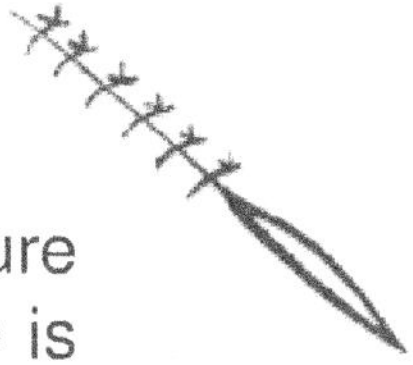

1. Interrupted

Interrupted suture has a separate and independent unit of suture and is easy to apply but is time consuming. This type of suture is generally used to suture the skin.

2. Continuous

Continuous suture does not have separate units of suture and requires less time to apply in the tissue, but a break at any point along the suture line results in total disruption of the suture.

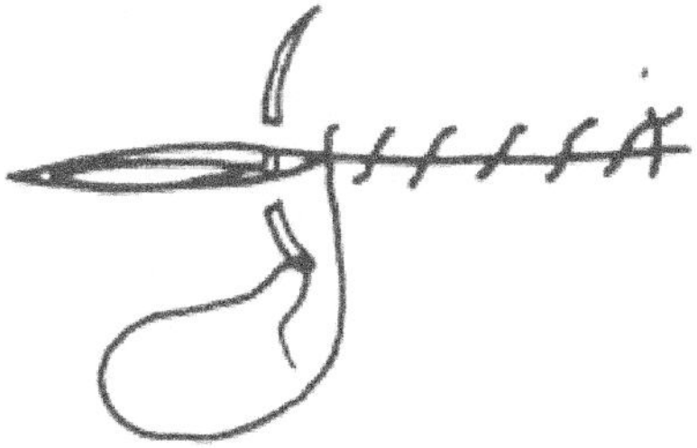

Method of Application of Sutures

- Sterilize the surgical instruments
- Restrain the animal properly
- Clean the wound according to its type, i.e. fresh, infected etc.
- Sterilize the wound, then your hands
- Use the absorbable suture material (catgut) to suture the internal part (muscle) and cotton/nylon for the external part (skin)
- Apply antibiotic powder on the wound and use intramuscular injections of antibiotics, if required
- Dress the wound daily or on alternating days
- Remove the sutures in the skin after completion of the healing process (usually after one or two weeks)

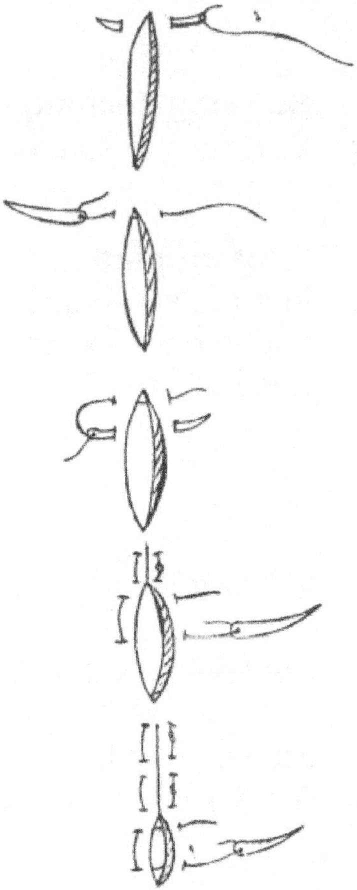

Type of Interrupted Suture (Figure)

Fractures

What is Fracture?

A fracture is defined as a break in the continuity of bone, with or without displacement of fragments. Fractures are associated with soft tissue damage of varying degree and obstruction in blood circulation. This problem is very common in hilly areas and extreme terrains.

Types of Fractures

1. Simple Fracture (Closed Fracture)

There is a fracture in the bone but no wound at the surface.

2. Compound Fracture (Open Fracture)

There is a fracture with an open wound extending through the skin down to the bone injury.

3. Comminuted Fracture

There is a fracture with multiple bone fragments. The bone may or may not be broken right through. This fracture may or may not be open.

Symptoms of Fracture

- Deformity at the site of fracture
- Loss of function
- Abnormal mobility
- Pain at the site of fracture
- Crepitous (rubbing) sound at the fracture site

Treatment of a Fracture

The treatment of a fracture depends on the location of the fracture and its complexity. The main principle in treatment of a fracture is the reduction of fractured fragments in its position and its immobilization. The following points should be taken into consideration for treatment of a fracture.

- Restrain the animal properly
- Avoid unnecessary handling of the patient
- Control the bleeding
- Remove all the contaminated and devitalized tissues
- Treat the wound, if any
- Reduction of fractured fragments (into their normal position)
- Retention and immobilization by using a splint or cast

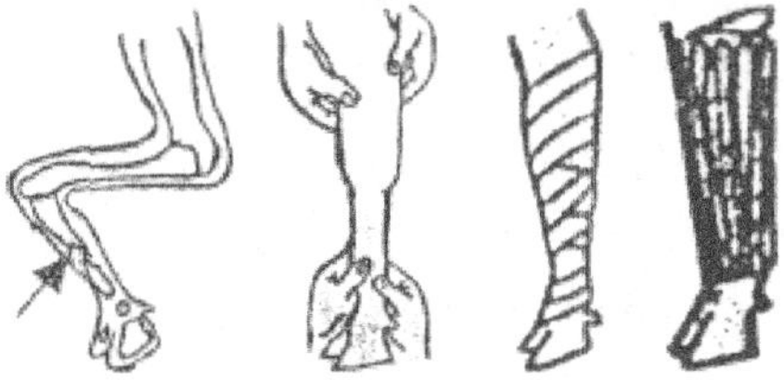

Bamboo/Wooden Splint

A bamboo splint is generally used to immobilize the fracture segments, especially in rural areas of Nepal. The method of preparing a bamboo splint is as follows:

- Collect 7-8 pieces of bamboo
- Tie the pieces with thread/thin jute rope on both the ends, leaving 5 inches from the ends of the bamboo pieces as shown in the figure on the right.
- While tying the ends, care should be taken to make one end narrower than the other, which makes it easier to immobilize the fractures, especially of legs.

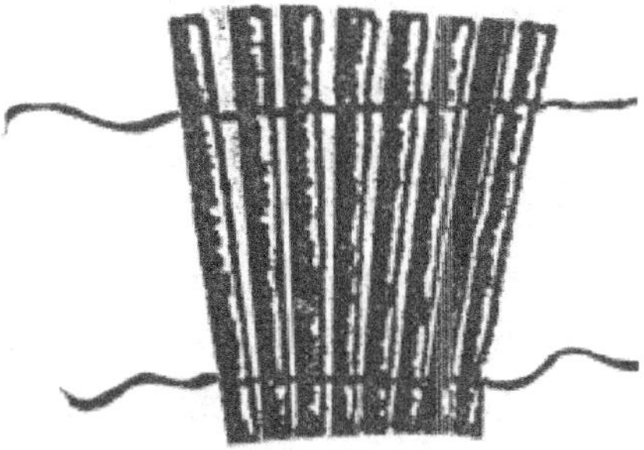

Abscess

What is an abscess?

- A wound under the skin that has become infected and full of pus.
- There is no visible opening, but a swelling can be seen.
- Usually found over the muscular parts of the body.

How do we recognise an abscess?

- There is a swollen lump on the animal.
- The swollen part will be red, hot and soft if it is a few days old. It may be cold and hard if it has been present for a long time.
- It is usually painful.
- Animal may become thin and weak if not treated promptly.

How do we know if the abscess is ready to drain?

- We can feel a soft part where the skin is getting thin.
- We can insert a 16 gauge needle into the abscess. If it is ready, pus will come out.
- If we are still unsure, we can connect a syringe to the needle to try to remove some of the pus.
- If it is ready, pus will come out. Pus is yellow or white and smells bad.
- If it is not ready, then only blood or water will come out.

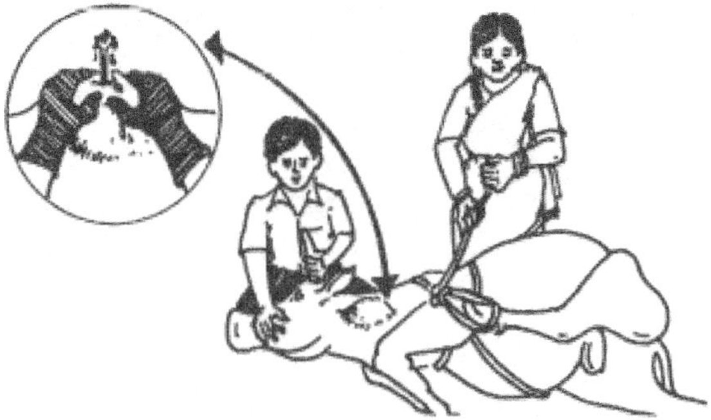

- Treatment (draining) should only be attempted if the abscess is ready.
- If it is not ready, we can wait a few more days.
- To speed up the process, we can soak the abscess each day with a warm, wet compress using water and magnesium sulphate.

How do we treat an abscess?

- Sterilize a shaving blade with a flame or by soaking in antiseptic solution.
- Using the sterilized blade, make an opening into the abscess.
- The cut should be straight.
- The opening must be made big enough to put 2 fingers through (about 3 cm).

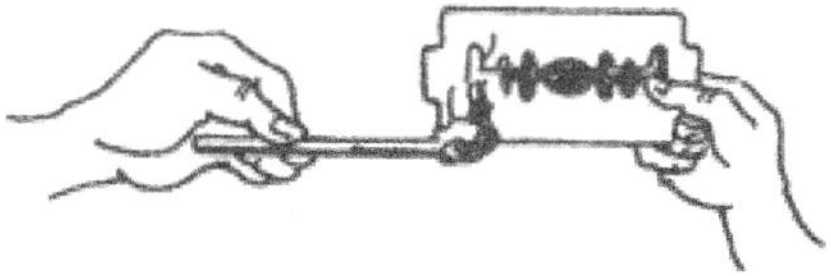

- Gently press on the wound to remove all the pus. Flush the wound with antiseptic solution using a syringe (without a needle attached). Dry the wound.
- Dip a small piece of cotton cloth in iodine tincture and stuff into the abscess cavity. This will keep the wound open so all the pus can come out.

- After a day, remove the cloth. Clean and dry the wound and apply Himax® ointment.
- Clean the wound and apply ointment each day until it is healed.
- If necessary, give penicillin injections.

Some Problems related to the Eye

Just like any other part of the body, the eye can also have different kinds of problems, such as:

- Eye infections
- Cloudy or white cornea (the *cornea* is the transparent part of the eyeball)
- Wounds

There are many kinds of home remedies used for treating eye problems, but you must be careful! Some things are not safe to use in the eyes!

1. Eye Infections:

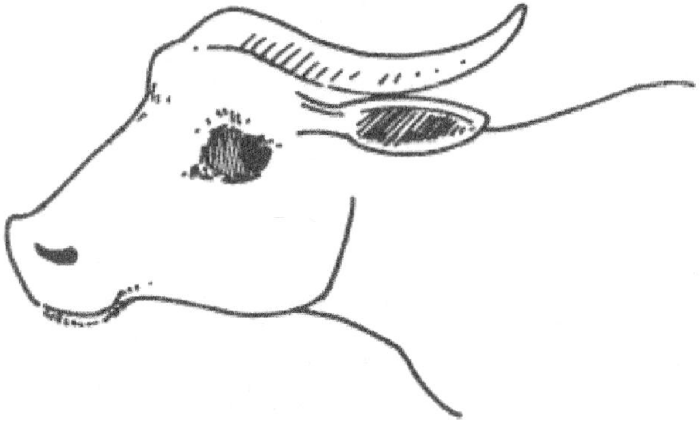

These infections are caused by *microscopic organisms.* Microscopic organisms cannot be seen with the naked eye. Some signs of an eye infection are:

- An infected eye has a lot of discharge.
- The white part of the eye will become red and swollen.
- The eye will have tears continuously.
- The eyelid may become swollen.
- The animal will blink in sunlight because the light causes pain.
- 'Pink eye' is a common eye infection, a small pink ulcer is found on the centre of the eye.

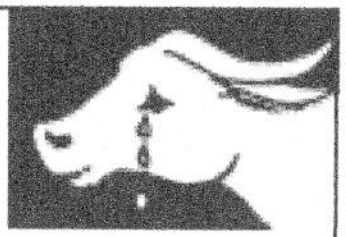

Treatment:

- Wash the eye gently with a boric acid solution or lukewarm water.
- Use Terramycin® eye ointment 3 times daily until healed (apply like in picture on page 107).
- If eye ointment does not clear the infection, injectable antibiotics can be used.

Prevention:

- Keep the affected animal away from the other animals, since these infections can pass to other animals.

2. Cloudy or white cornea

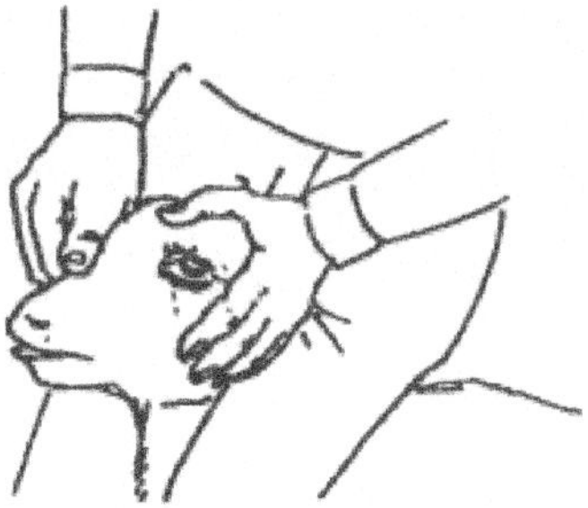

Sometimes the clear, front part of the eye will become cloudy or white colored. This *cloudy cornea* may be only a small spot or may cover the entire cornea. If treatment is delayed, the white colour change may be permanent. There are many causes of white cornea, but the most common is injury from a stick or other kind of wound.

Treatment:

- Wash the eye gently with a boric acid solution or lukewarm water.
- Apply alum powder to the eye as shown. Other items available to treat white cornea are whey and turmeric powder.

3. Eye wounds:

- Grass may penetrate the eye and cause a wound.
- Animals with eye wounds will have lots of discharge from the eye.
- Large wounds can become infected, which can result in permanent blindness.

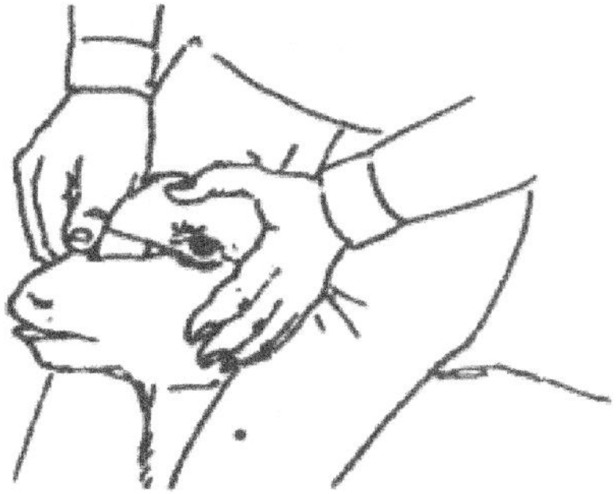

Treatment:

- Wash the eye gently with a boric acid solution or lukewarm water.
- Use Terramycin® eye ointment three times per day.

Liver Fluke

Liver fluke is a problem in all 14 zones and 75 districts in Nepal. It is known by various local names, such as: *kaleje, kalejoko kira, leu, paani laagne,chiso parne, chamre,* and *chyaapte kiraa* etc.

What is it?

- It is an internal parasite
- It lives in the liver of the animal
- It looks like a flat leaf, dark red in color.

How does an animal get liver fluke?

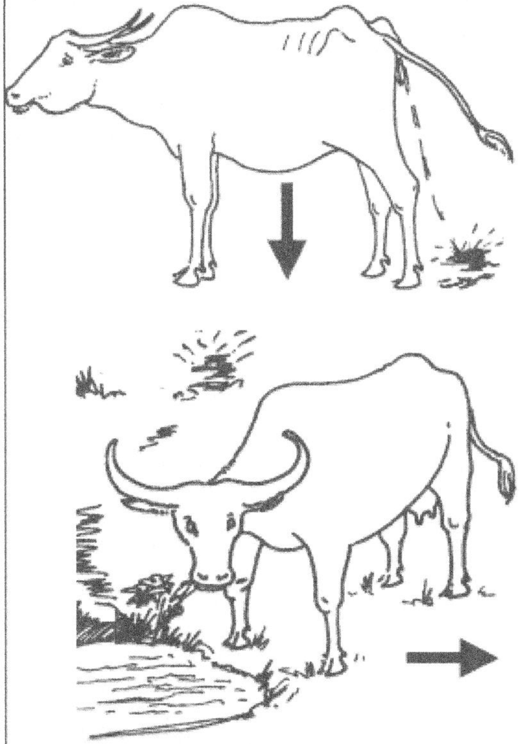

Liver fluke eggs are passed out in the faeces of affected animals. *Larvae* (young flukes) are hatched out from these eggs. The larvae enter the body of a snail. After a few days of growth, the larvae come out of the snail and attach to the grass or a rice plant. When the animal eats the grass or wet straw, it also eats the liver fluke larvae.

These larvae then pass through the stomach and intestines and eventually settle in the liver. Then they begin to damage the liver and eat the bile, the signs of the disease begin to appear.

Which animals are affected by liver fluke?

- Commonly raised livestock such as: cow, buffalo, sheep, and goat.
- It is a major problem in cows and buffaloes.
- Nursing calves, kids, and lambs are not affected.

What does an animal look like that has liver fluke?

- Slowly becomes thin.
- Does not grow properly.
- Hollow, sunken appearance around pelvis.
- Has diarrhoea.
- Licks the soil.
- If the animal becomes too thin, the chin may become swollen.
- The *conjunctiva* (area around eyeball) becomes pale.
- At the end the animal becomes very thin and may die.
- Does not show signs of heat in time.

What treatments are available?

Many different preparations are available for the treatment of liver fluke.

For example: Distodin® (oxyclozanide), Tolzan®, Tetzan®, Valbazen®, Hexanide®, Clozan®, Wokazan®, and CTC (carbon tetrachloride).

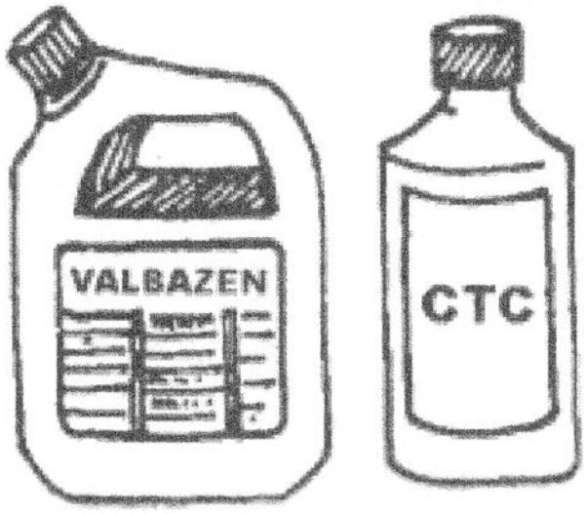

Method of using liver fluke medicine

Give the full dose to healthy and strong animals

For sick and thin animals:

- Give vitamin B injections for 3 days.
- On the 4th day, give a half dose of the medicine to the animal.
- On the 6th day, give the second half of the dose of the medicine.

Prevention:

1. Healthy animals should be given liver fluke medicine every 6 months.

Treatment for calves should start when they are 1 year old. Treatment for kids and lambs should be started at 5 months of age.

2. Stall-feeding

Stall-fed animals have fewer problems with liver fluke. Grazing in wet, marshy areas leads to more consumption of liver fluke larvae.

Large Round Worms

What are large round worms?

- They are internal parasites.
- They are found in the intestine and stomach.
- They eat the nutrients in the food that the animal needs.
- They appear long, round, and white .
- Worms of different lengths affect different kinds of animals.

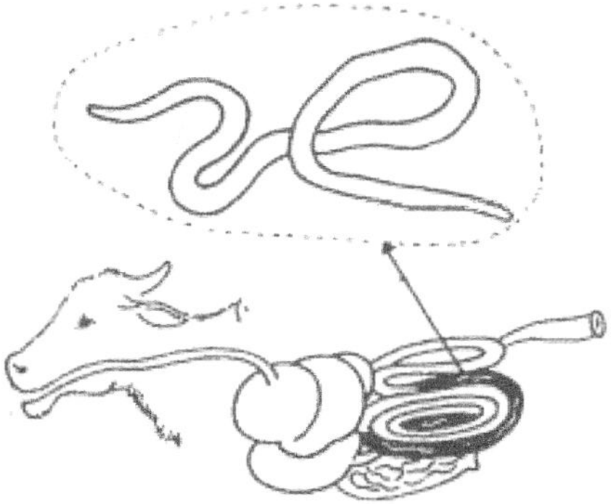

Which animals are affected by large roundworms?

- Non-ruminant animals of all ages such as: horses, pigs, dogs, cats, and chickens
- Nursing calves.

How does an animal get large round worms?

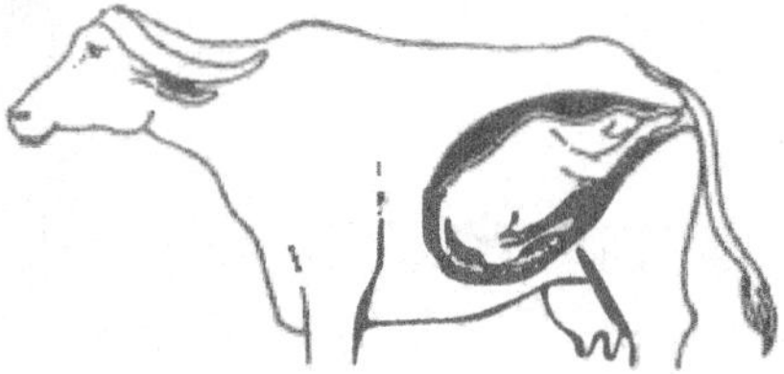

- They can be infected inside the uterus (before they are born).
- The eggs can be eaten while sucking milk from the mother.
- By eating contaminated grass or feed.

What do affected animals look like?

- Poor growth.
- Weight loss.
- Diarrhoea.
- Constipation if large numbers of worms are present.
- Pale conjunctiva (area around the eyeball).
- Sleep lying down (weak).
- Bad odor to the breath.
- Can sometimes see worms in the faeces.
- Stomach may look bloated.
- Dogs, cats, and pigs may vomit worms.
- Appetite may be poor.

How do we treat it?

1. Feed piperazine or Piperacite®.
Dose rate depends on the type of animal. You can find the dose rate in your medicine book.

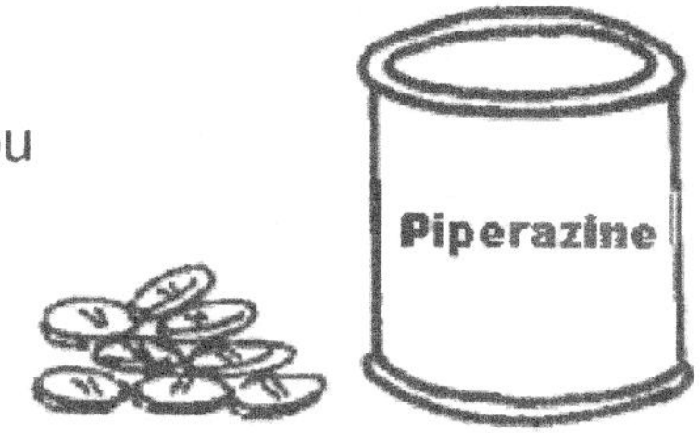

Prevention and Control:

- Give dewormer (piperazine) regularly.
- Give only clean feed and clean grass to animals.
- Shed should be clean and dry.
- Feed only clean milk to calves.
- Clean udder of nursing dam regularly.

Small Roundworms

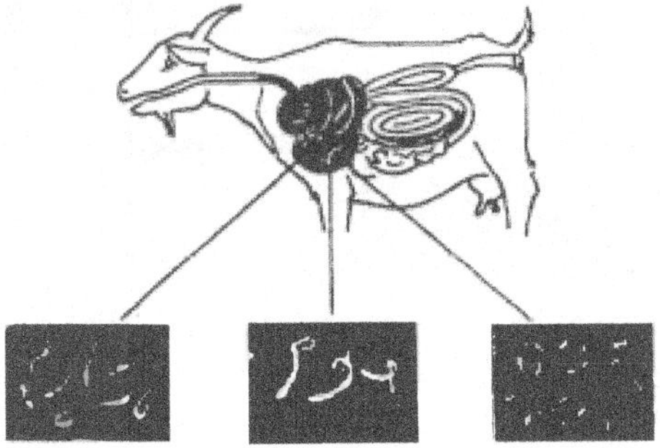

What are small roundworms?

- They are internal parasites.
- They are found in the stomach, lungs and small intestine.
- They suck blood from the animals.
- They are always small and round, but come in many shapes and sizes.

Which animals are affected?

- Small roundworms affect all animals.
- They are mainly a problem for goats and sheep.

What do affected animals look like?

- Gradually become thin.
- Sometimes have diarrhoea .
- Sometimes cough frequently.

- Poor growth.
- Hair loss in some animals.
- May not come into heat.

- Pale conjunctiva.
- Poor appetite.
- Appear depressed and hold their head down.

How do animals get small round worms?

Eggs are passed in the faeces of affected animals. These eggs stick to the grass. When healthy animals eat the grass, the eggs reach the animal's stomach. The eggs then hatch and the larvae (young worms) attach themselves to the intestine and begin to suck blood. The signs will appear if there are large numbers of worms inside the animal.

What treatment is available?

There are many types of medication available to treat small roundworms.

For example: Panacur®, Panfugal®, Valbazen®

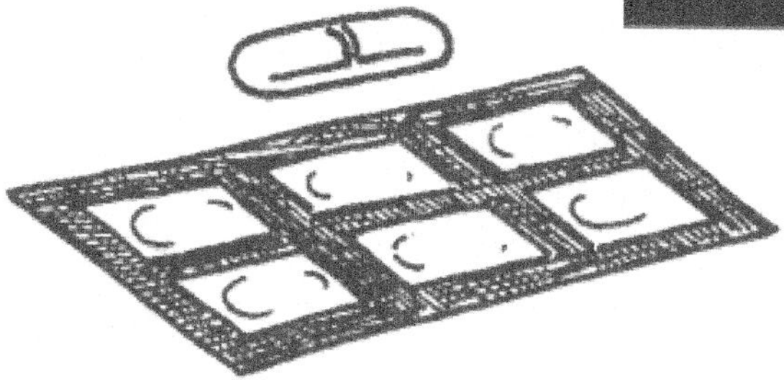

Prevention and Control:

1. Place feed in a manger (off the ground). If a manger is not available, tie the grass into a bundle and hang from the roof of the shed.

2. Keep goats in a shed with a slatted floor so that the feed will not become contaminated with faeces.

Lice

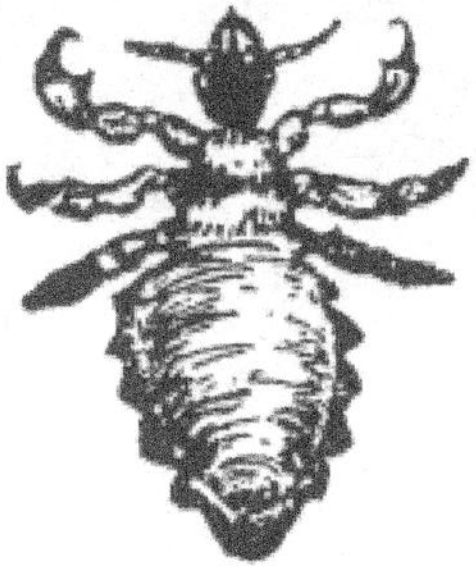

Buffalo Louse

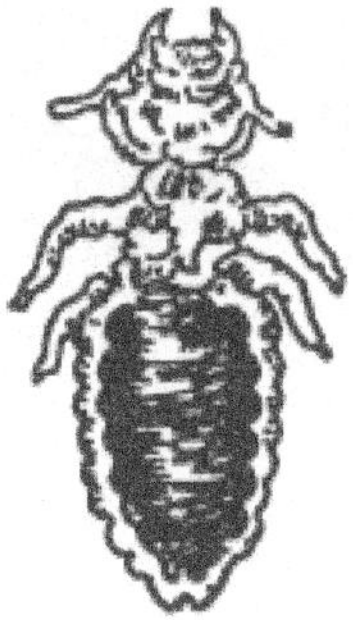

Sheep Louse

What are lice?

- They are external parasites.
- They are brown or black in colour.
- Size, shape, and colour vary with the kinds of animals affected.
- They live on the skin or in the hair.
- They live by eating skin or sucking blood.
- They lay white eggs (nits) on the hair.
- They are more common on thin, weak animals.

Which animals are affected by lice?

All animals including cows, buffaloes, sheep. goats, horses, pigs, and poultry.

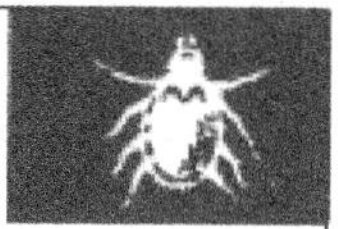

How do we recognise lice?

- Affected animals are often thin.
- Hair coat is rough and thin.
- Lice and eggs can be seen on the hair and skin.
- Affected animals scratch frequently.

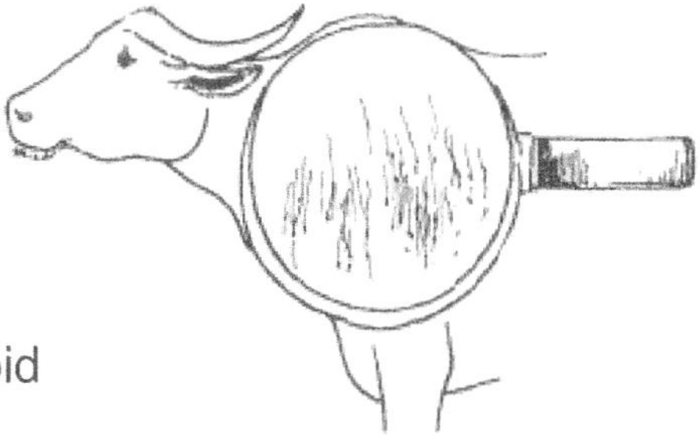

Prevention:

- Do not place too many animals in one shed—avoid crowding.
- Keep shed clean and allow sunlight and fresh air into shed.
- Keep animals clean.

Treatment:

- Butox®
- Neocidol®

Home Treatment:

- Rub animal's skin with tobacco solution.
- Apply once a week.

Fleas

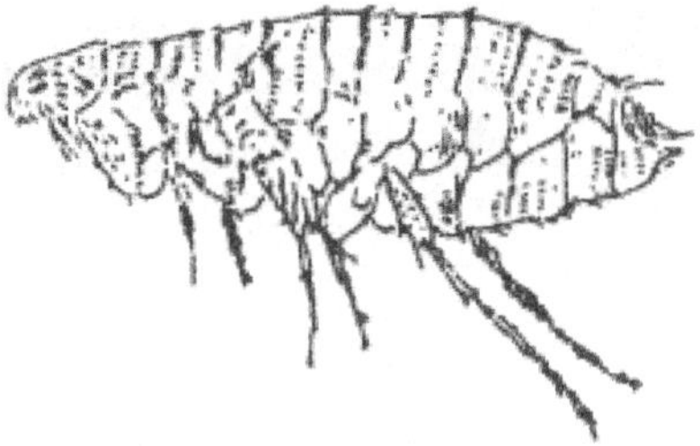

What are fleas?

- They are external parasites.
- They suck blood.
- They live in dark places.
- Eggs laid by the flea roll off the animal and hatch in the dirt.
- They transmit disease.

Which animals are affected by fleas?

All animals can be infested with fleas, but animals with especially thin or thick hair coats are mostly affected.

For example: Sheep, goats, dog, cat, poultry

How do we recognise an animal with fleas?

- Fleas can be seen on the animal by parting the hair.
- Areas bitten by the flea are very itchy.
- Areas bitten by the flea are red and swollen.
- Flea faeces can be seen on the skin. They look like black powder but are really digested blood. If the faeces are mixed with water, they will become red.

Treatment:

1. Use Butox® or Neocidol® on animal as directed and spray medicine into dark areas of shed.

Or

2. Rub tobacco solution on animal once a week.

Prevention and Control

- Sheds should be built where sunlight reaches all the corners.
- Keep the shed and the area around the shed clean.

- Kill mice and rats because they carry fleas.

- Change the bedding regularly.
- Burn old bedding.

Ticks

What are ticks?

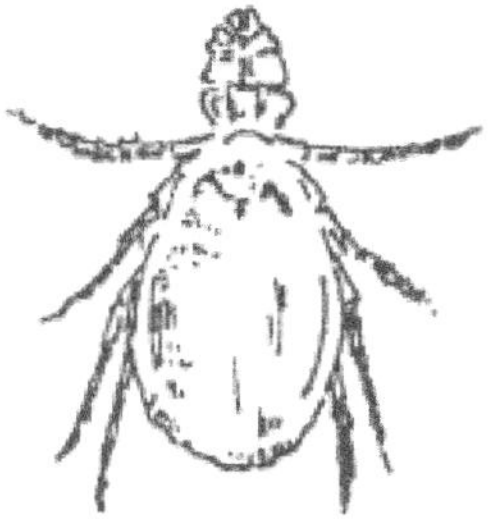

- They are external parasites.
- They may be black, brown, red, or white in colour.
- They are found on the hairless or smooth areas of the body.
- They suck blood.
- After sucking enough blood, they will drop off the animal by themselves.

Which animals are affected by ticks?

Cattle, buffaloes, sheep, goats, dogs, and horses

How do we recognise a tick infestation?

- Ticks can be seen on the hairless parts of the body such as the base of the tail, the udder, around the ears, and under the legs.
- The area where the tick bites may be red and swollen.

Treatment:

1. Butox®
2. Neocidol®
3. If the ticks are few in number, they can be removed and killed by hand.

Rubbing salt on the ticks to remove them is commonly used but this treatment does not kill the tick.

Prevention:

- Stall-feeding. Ticks live in grassy, forest, and jungle areas. If animals are stall fed, it may reduce their contact with ticks.

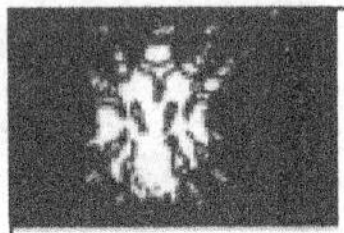

Mange

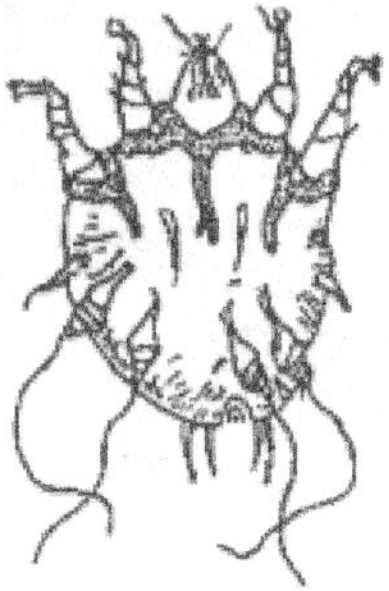

Mange is caused by microscopic mites *(microscopic* means unable to see with the eye alone). These mites burrow into the skin from the roots of the hairs. Slowly they spread over the whole skin.

Which animals are affected by mange?

Cattle, buffaloes, sheep, goats, pigs, dogs, and horses. Buffaloes and pigs are the most affected.

How do we recognise mange?

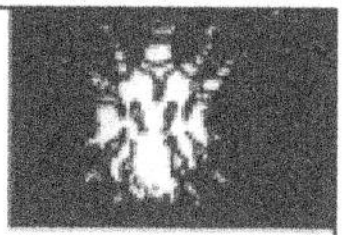

- Rough, thick, and scaly skin.
- The hair falls out and patches of bare skin can be seen.
- The hair comes out easily when pulled.
- The animal bites and scratches at his skin because it is very itchy.
- Wounds result from rubbing the skin. Later, these wounds can become infected.

Treatment:

Sulphur and mustard oil

- 10 parts mustard oil
- 1 part sulphur powder

Mix these together well and rub the mixture on the body of the animal.

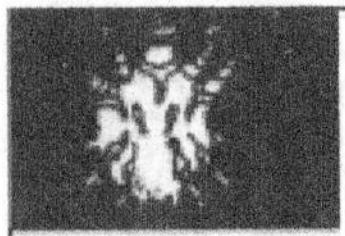

Other treatments:

1. Use Butox® or Neocidol®.
2. Himax® ointment applied to affected areas.
3. Solutions of *titepati* or *neem*.
4. Old motor oil can also be used.

Prevention:
- Keep the animals and the shed clean.

Caution :
Care must be taken as this disease can transmit from animal to man

Reproduction In Animals

Mating age

The age of sexual maturity (when the animal can be mated) varies between different species.

Local cow: 2 years
Improved-breed cow with proper nutrition: 12-18 months of age.

Sheep: 6 to 12 months

Buffalo: 2 1/2 years

Rabbit: 4 to 6 months

Horse: 18 to 24 months

Pig: 5 to 8 months

Goat: 7 months

Dog: 6 months

Pregnancy and Delivery Problems

- Keep the animal in a clean, dry, and well-ventilated shed.
- Feed the animal nutritious food.
- Give plenty of water.
- Feed the animal often, but give small portions rather than feeding a lot of food at one time.

Stages of Parturition

There are 3 stages of parturition *(parturition* means delivery of the young)

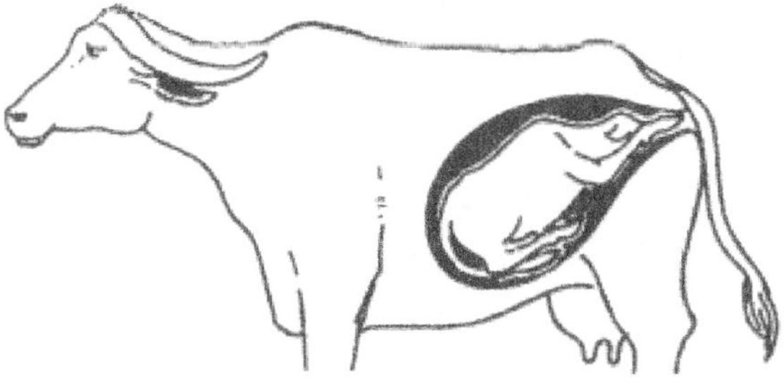

1. First stage:

- The animal loses appetite during this period.
- The animal will be restless and will continuously lie down and get up.
- The animal will start dripping urine.
- The animal will start straining.

2. Second stage:
At this stage water flows out of the uterus. In the cow, buffalo, sheep and goat, this stage will last 2 to 5 hours. It will last 1 to 4 hours in the horse and it will last 2 to 12 hours in the pig and dog. At the end of this stage the *fetus* (baby) comes out. Most animals give birth lying down.

3. Third stage:

The *placenta* (afterbirth) will normally pass within 1 to 8 hours after birth.

Dystocia

What is dystocia?

Dystocia is when the mother cannot deliver the baby on her own. Sometimes we must provide help.

First we must know about the *gestation period* of various animals. The gestation period is the time it takes for the baby to develop inside the mother.

Gestation periods are different for different kinds of animals.

Cow: 280 days

Sheep: 150 days

Buffalo: 310 days

Rabbit: 28 days

Horse: 335 days

Pig: 114 days

Goat: 150 days

Dog: 63 days

Causes of dystocia:

1. Weakness. Lack of nutritious feed and water can make an animal weak. Diseases can also make her weak so that she cannot push the baby out.
2. The cervix may not be open.
3. The uterus may be twisted.
4. The fetus may be in an abnormal position.

There are some questions that we should ask the owner before we decide to help at delivery time.

Questions to ask:

- When is the due date?
- When did labor start?
- Were there similar problems during previous deliveries?
- Has any one assisted in the delivery yet?
- Has the animal been in labor (pushing hard) more than 2 or 3 hours?
- If no one has yet tried to assist the delivery, then we should try to help by first putting our hand inside and seeing what we can feel.

Before putting your hand inside the uterus:

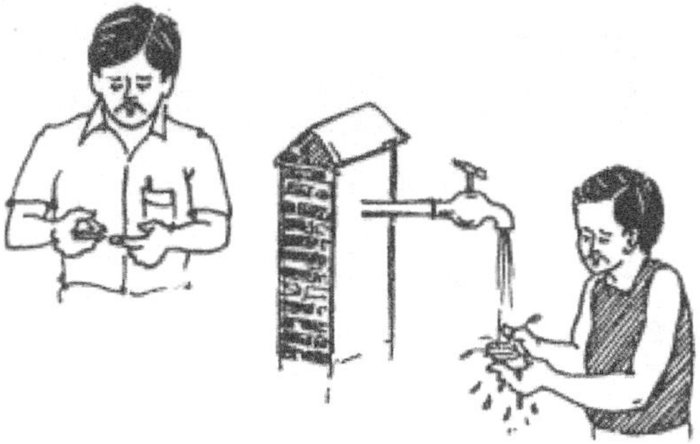

- Clean the shed where you will be working.
- Wash the rear of the mother with soap and water.
- Clip fingernails short and wash your hands.
- Make your hand and arm slippery with soap and water.
- If the uterus is dry, put some soapy water inside to make it slippery.
- Slide your hand and arm gently inside the uterus.

Should we assist?

You may find one of the following conditions inside the uterus

a. Closed cervix
b. Uterine torsion (twist)
c. Any part of the fetus

a) Closed cervix:

It may not be time for delivery. Check again after 6-8 hours during this time gentle walking may help. By then the cervix may have opened by itself. If it does open, wait a few more hours because the mother may be able to deliver the baby by herself.

b. Uterine Torsion (Twist)

In case of uterine torsion, when you place your arm in the birth canal you will find your arm bends. The birth canal seems narrow and twisted, this prevents you from entering into the uterus. This condition is hard to treat - veterinary assistance is required.

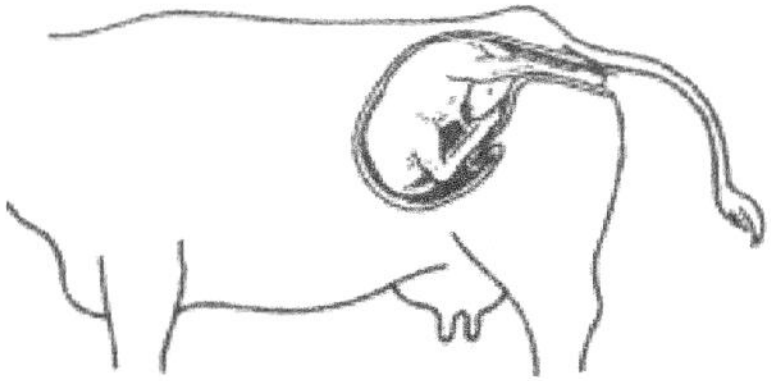

c) Any part of the fetus

- You may find a leg of the fetus.
- If so, you must determine if it is a hind leg or a front leg.

Front leg or back leg?

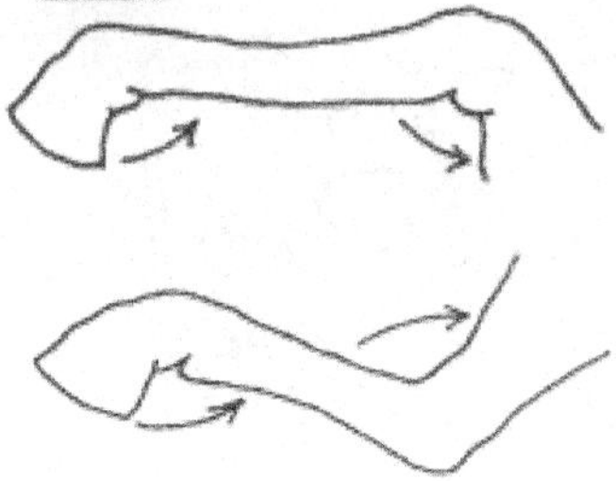

- On the front leg, both hoof joint and the upper joint *(carpus)* bend in the same direction.
- On the hind leg, the hoof joint and the upper joint (tarsus) bend in opposite directions.
- The long bone of the front leg is round, but the rear leg is more flat.

After determining which leg is showing, you can try to re-position the fetus for delivery.

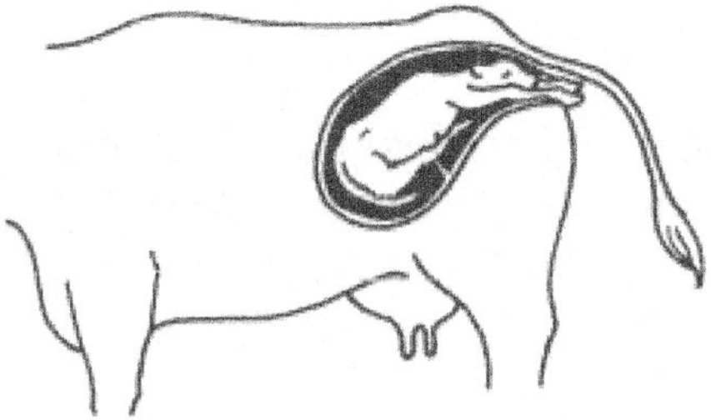

What are the normal positions for delivery?

1. With the 2 front legs straight and the head between them the animal can deliver normally.

or

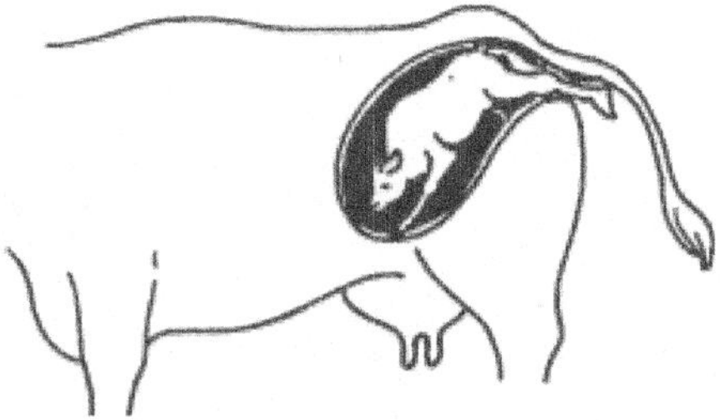

2. The animal can also deliver the baby with the 2 rear legs first.

Any other position besides these two means that the animal will not be able to deliver the baby without help.

Treatment of various incorrect positions

General Consideration:

Before correcting the problem in the following examples, you must first gently push the baby farther into the uterus to gain more space for maneuvering. **Always try to bring the position of fetus to the normal position.**

Caution

- Only push the baby back into the uterus when the mother is not pushing. Once the baby is delivered, antibiotics should be given to the mother to prevent uterine infection.
- Never try to pull the baby out of its mother before the baby is correctly positioned.

1. The head and only one leg are forward:

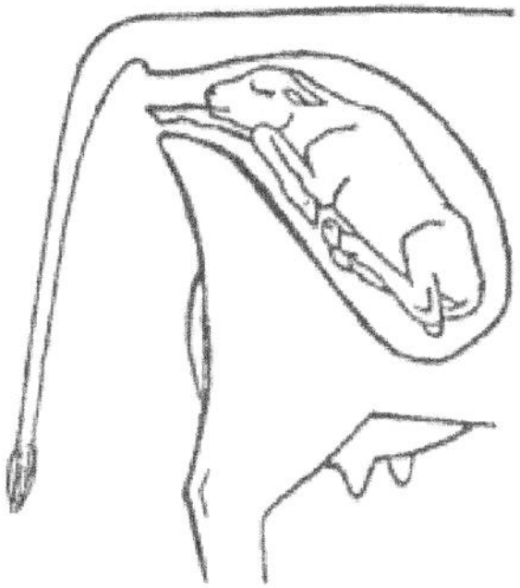

- Attach a rope to the front leg that is forward, but do not pull it.
- Slide your hand down the opposite side of the animal and find the leg that is back.
- Cupping your hands around the hoof edges to avoid tearing the uterus, carefully move the leg into the correct position.
- Once in the correct position, pull the baby out.

2. Only head is forward and both front legs are back

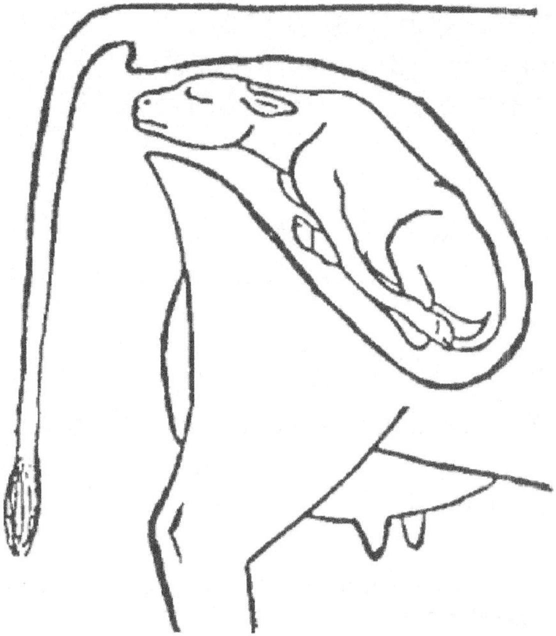

- Slide your hand down each side of the baby and grab the legs.
- Cupping your hands around the hoof edges to avoid tearing the uterus, carefully move each leg into the correct position.
- Once in the correct position, pull the baby out. Attach ropes to the front legs if necessary.

3. Front legs only are forward and the head is turned back

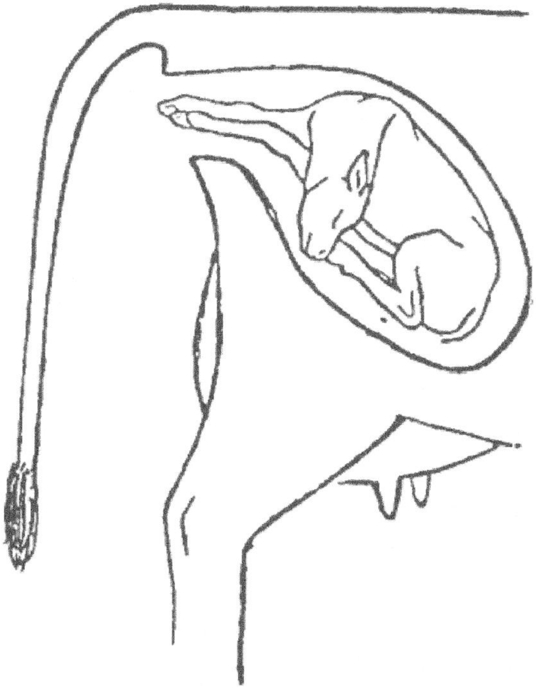

- Attach ropes to the front legs but do not pull on the ropes.
- Search for the head.
- After finding the head, grab the mouth, eyes sockets or ears and gently pull into the correct position. It may be necessary to untwist the neck first.

Caution : Cup your hands around the baby's teeth when pulling the head forward to avoid tearing the uterus with the baby's teeth.

- Once in the correct position, pull on the ropes to pull the baby out.

4. The tail only is present without the back legs (Breech Position)

Sometimes only the tail is found and both legs are tucked up under the body.

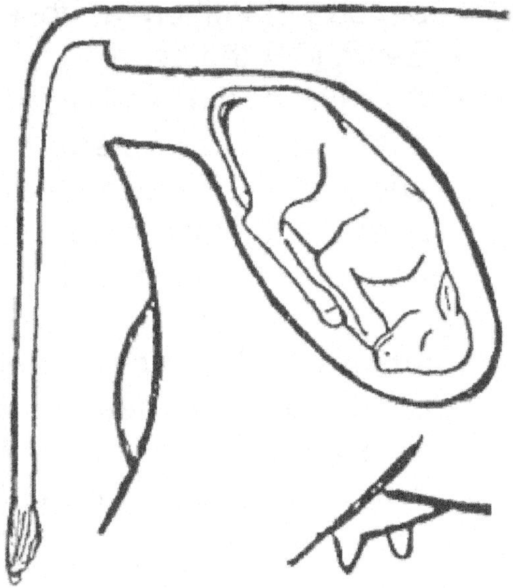

- Put your hand on the rump of the baby and gently push it back in the birth canal so that you have room to correct the position of the back legs. Do this carefully, when the mother is not pushing.
- Reach in and find the hock on one back leg and pull it toward you and up. Then place second hand over hoof, flexing it back and slightly to the center while you pull it into the birth canal.
- At the same time you may need to push up and in on the hock to allow room for the leg to flex and hoof to be pulled back. Your hand over the hoof will protect the hoof from tearing the uterus.
- Do the same for the other leg until both back legs and tail are in the birth canal.
- Pull on the legs until the baby comes out. It may be necessary to attach ropes to the legs.

Retained Placenta

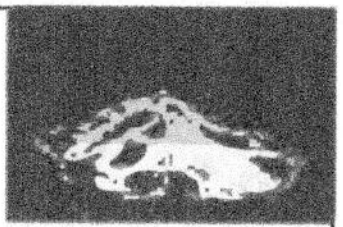

What is a retained placenta?
Normally 1 to 8 hours after birth the *placenta* (afterbirth) will pass on its own. If it does not, it is called a *retained placenta*. To encourage the placenta to pass, allow the calf to nurse.

Causes of a retained placenta:

- Dystocia
- Weakness of the mother
- Premature birth or abortion
- Birth of twins
- Birth of dead fetus

How do we treat a retained placenta?

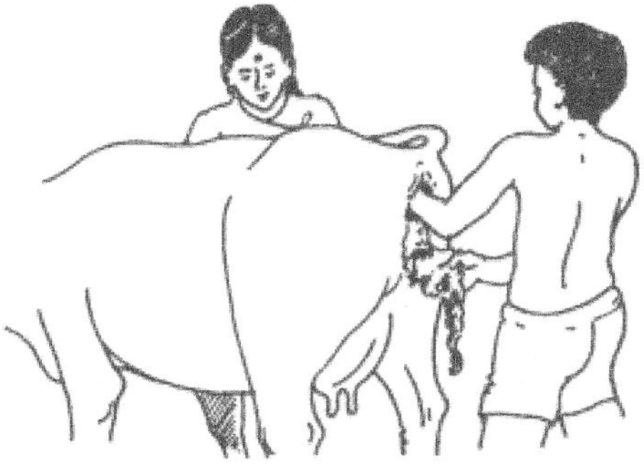

If the retained placenta is not treated in time, it can cause problems for the next pregnancy. For example, if the uterus becomes infected it can lead to infertility.

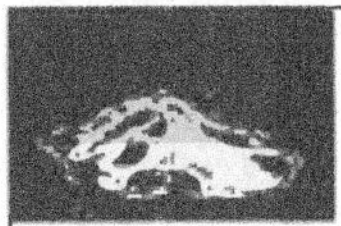

Treatment:

Give Exapar liquid.

- 100 ml. orally on first day, then 50 ml. orally for 3-5 days

Other treatments:

- If there is a fever, give penicillin or Terramycin® injections.
- Clean the animal with Dettol® or Savlon® solution and place 2 sulfadimidine tablets into the uterus.

Prevention:

- Allow the calf to nurse immediately after delivery.
- Give Exapar®.

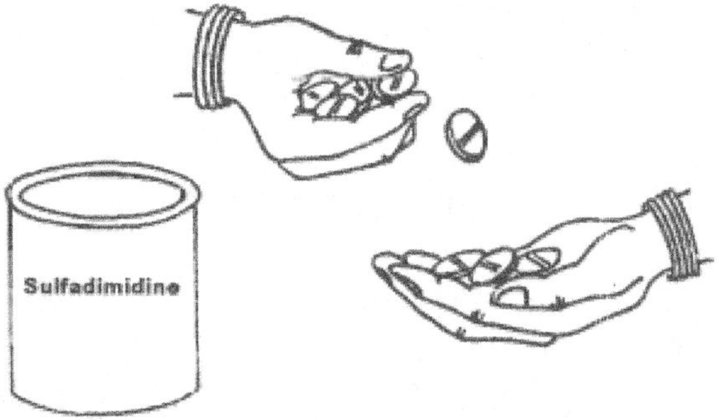

Uterine Prolapse

What is a uterine prolapse?

After the delivery of the baby, the mother may continue to strain. This may cause the uterus to come outside the body. This is known as a *uterine prolapse.* This is most commonly seen in the cow and buffalo, but can occur in any kind of animal.

- This occurs within 1 to 2 days of delivery
- The animal may be restless and strain continuously
- The uterus can be seen coming out of the vulva
- It may look like a large amount of red meat
- If treatment is not given immediately the animal will die.

Causes of uterine prolapse:

- The calf does not nurse immediately after birth.
- Weakness of the uterus muscles.
- Poor nutrition.
- Eating moldy or rotten feed.
- Over feeding of green grasses and young leaves.
- Eating large amounts of salt all at one time.
- Retained placenta.
- Injury to the uterus during delivery.
- Pain in the uterus from infection.
- Excessive straining because of diarrhoea.
- Keeping the animal in a shed with a floor that slopes down to the back.

Treatment:

- Clip fingernails and wash your hands before handling the uterus.
- If the animal is not weak, make it stand up.
- If the animal is weak, place the animal with the front end a little lower than the back end. Pull the hind legs straight back.

- Place the uterus in a clean plastic sheet or cloth.
- Wash the uterus gently with warm water and antiseptic solution.
- If it is very swollen, we can apply citrus juice, sugar, or magnesium sulphate powder to shrink it.
- If the animal tries to rise when it is laying down, it may help to have somebody sit on it's lower back.

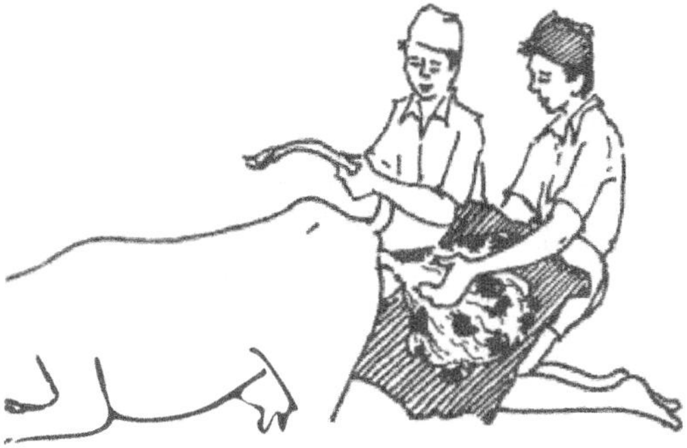

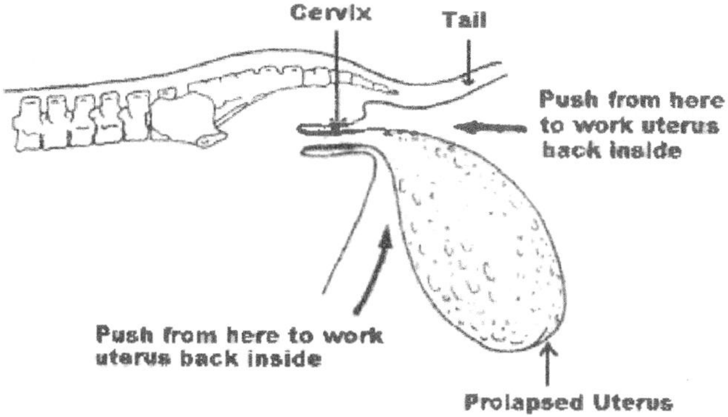

- Place the uterus on your knees and push the uterus back inside with both hands. Gradually work all the tissues back inside.

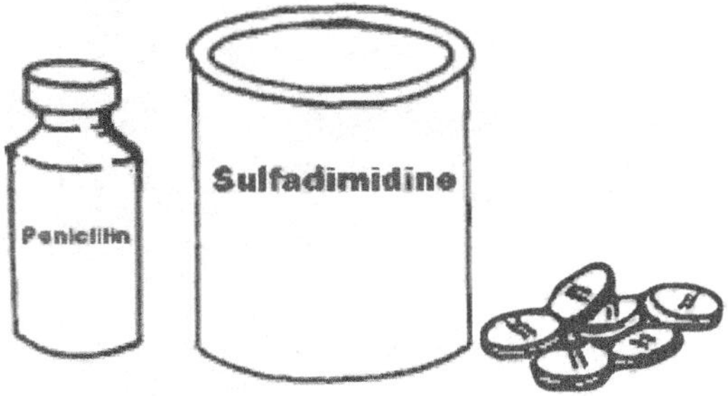

- You must push only when the animal stops straining.
- The uterus must be returned fully to its normal position. If your arm is not long enough to fully replace the uterus, use the bottom of a clean bottle to push the last part inside.
- Flushing the uterus using an antiseptic solution can also help.
- When finished, place 2 sulfadimidine tablets inside the uterus.
- Give penicillin injections.

- Tie the animal so that its front end is a little lower than the back end.
- Sometimes a ring like the one shown on the right can help keep the uterus inside. Place the ring around the outside of the birth canal, with the upper two ropes running a long the animal's back, and the lower two ropes running along the animal's abdomen, then tie these ropes to a neck rope.
- You can also stitch the vulva (mouth of the birth canal) closed. Use clean cotton or plastic string and remove it in a few days.

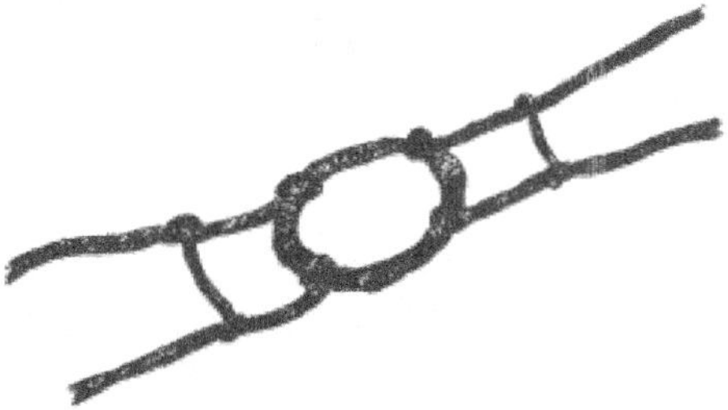

Prevention

- Keep animals in good, healthy condition. Give medication regularly for worms and liver fluke. Feed a balanced and nutritious diet.
- Do not give more than 50 grams of salt per day.
- Do not feed animals large quantities of water and food at delivery time. Give a little food frequently throughout the day.

- Do not feed mouldy or rotten feed.
- Keep pregnant animals on a slightly sloping floor, the front end a little lower than the back end.
- If the animal survives a prolapsed uterus, she should be sold for meat after the calf is weaned. Do not try to breed her again. Do not keep female calves from these problem animals.

Vaginal Prolapse

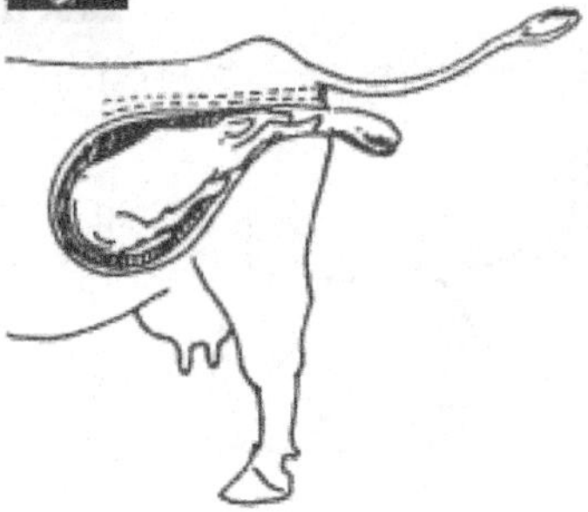

What is a vaginal prolapse?

Before delivery sometimes the vagina and cervix may come out of the vulva. This condition is called a *vaginal prolapse.*

Causes of a vaginal prolapse:

- The shed floor slopes to the back. This puts pressure on the uterus and cervix.
- Over feeding.
- Pain or infection in the uterus.
- Feeding mouldy or rotten feed.
- Giving large amounts of salt at one time.
- Some breeds of animals are more likely to get this because of the way their body is shaped.
- Hereditary problems.
- Lack of minerals.

Treatment

- As soon as you see the prolapse, clean it with warm water and antiseptic and make the animal stand up.
- Most of the time the vaginal prolapse will go in by itself. If it does not, you can try to replace it by hand.

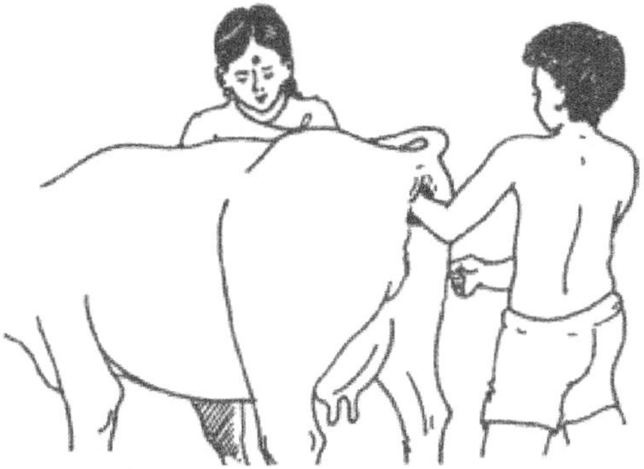

Prevention:

- Keep the animal in a shed with the floor sloping a little to the front.
- Give small amounts of feed throughout the day instead of giving it all at once.
- Do not feed mouldy or rotten feed.
- Do not give large amounts of salt at one time. Feed a little salt each day.

Care of the Newborn Animal

- Remove mucus, straw, and dirt from the mouth and nose of the newborn so it can breathe easily.
- Cut the umbilical cord about 3-5 cm away from the body. Iodine on the cord can be used to prevent infection.
- Let the mother lick the newborn as this will help bond the mother to the baby as well as help with milk production and shrinking of the uterus.
- Encourage the calf to suck milk as soon as possible. The colostrum (first milk) is beneficial to the calf. It is nutritious and helps the young animal fight off disease. Calves fed colostrum grow rapidly and are healthy.
- Feed the mother good quality grass and feed.
- Encourage the mother to get up and move around immediately after birth. This will help to prevent uterine prolapse.
- Clean the udder properly before and after milking.

Things to remember:

- Keep the shed clean and dry.
- Make the resting area in the shed lower in the front and higher in the back.
- Give clean and nutritious feed, but do not give too much at once.
- Give lots of water so the mother can make lots of milk.

Importance of Colostrum

What is colostrum?

The milk that comes on the first 3 days after birth is called *colostrum*.

What is in colostrum?

- The material required for good growth of bones and hair.
- The material required for the development of flesh and blood in the body.
- The material required to fight off different diseases.
- Nutrients that are easy for the newborn animal to digest.

What are the benefits of feeding colostrum?

- It will help the placenta (afterbirth) to come out on time.
- The newborn animal will have easy to digest food right away.
- There will be no problems with constipation compared to feeding milk.
- Rapid growth of the newborn animal.
- The newborn animal will be better able to fight off disease.
- Rapid maturity to reproductive age.
- Good production once the animal has matured.

How do we feed colostrum?

- Clean the teats and udder with warm water after the mother gives birth
- Encourage the newborn to suckle
- Feed enough colostrum, for the first 3 days to the newborn.
- If the mother will not allow the baby to suckle, then milk out the mother and feed the colostrum to the baby with a bottle or bamboo pipe. Feed 1 liter of colostrum per 10 kg body weight each day for 3 days.

Note: If hand-feeding the baby, utensils and colostrum must be clean!

Mastitis (Swollen Udder, Bad Milk)

What is it?

Mastitis means an inflammation (swelling) of the udder. This usually means that the udder or teats are swollen, hot, and reddened. Mastitis lowers milk production.

Causes

The common causes of mastitis are

- Bacterial infection
- Injury to the udder

Predisposing Factors

- Age - the disease is more prevalent in older animals.
- Stage of lactation – the risk of infection is greater in the initial and end stages of lactation.
- Complete milking - incomplete removal of milk from the udder may lead to mastitis
- Feed offered - heavy protein feeding may predispose the condition
- Genetic factors such as the confirmation of the udder and teats.
- Trauma - any injury to the teat or udder helps the microorganisms gain entry, often caused by poor milking technique.
- Hygiene - bad hygiene and sanitation

Which animals are affected?

- All breeds of cattle, buffalo, sheep, goat, pigs and horse are susceptible.
- High milk yielders are more commonly affected than lower yielders.
- Infection rate is more in successive lactations than the first.
- Exotic and crossbred animals are more prone to mastitis.

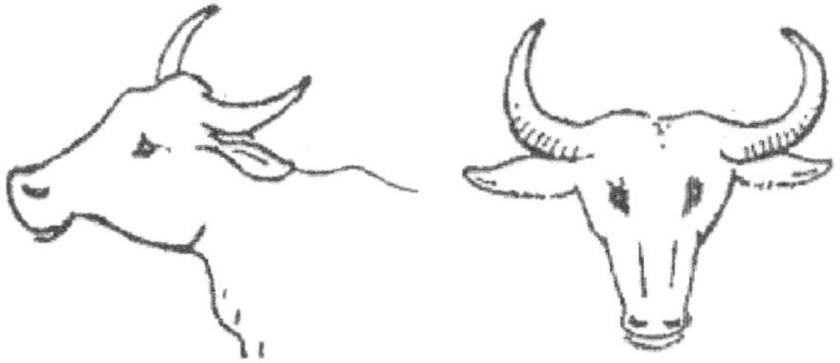

Symptoms

- The udder or teat may become swollen, hot and painful
- Lumps can be felt in the udder
- Milk may coagulate (clump)
- The taste of the milk may be bad
- The colour of milk may be different
- There may be blood or pus in the milk.
- Animals give less milk than usual
- If disease is not treated, the teat may drop off.
- The animal may have a fever.

Diagnosis

- Symptoms.
- Milk test in laboratory.

Treatment

- Milk frequently - milking frequently, at least 4 times daily, helps by decreasing the pressure and pain, while flushing out the udder as well. If no milk will come out, a clean teat canula should be inserted into the teat so that the milk can come out.

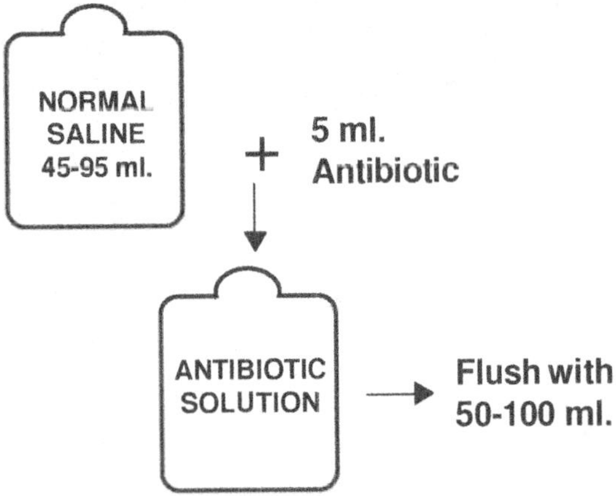

- Apply a warm, wet cloth to the udder and teat - it helps to reduce swelling and pain.
- Give antibiotics which are sensitive to the infection (after having a milk test done in the lab).
- Flushing of udder with an antibiotic and normal saline solution will be beneficial.
- Use intra mammary antibiotics

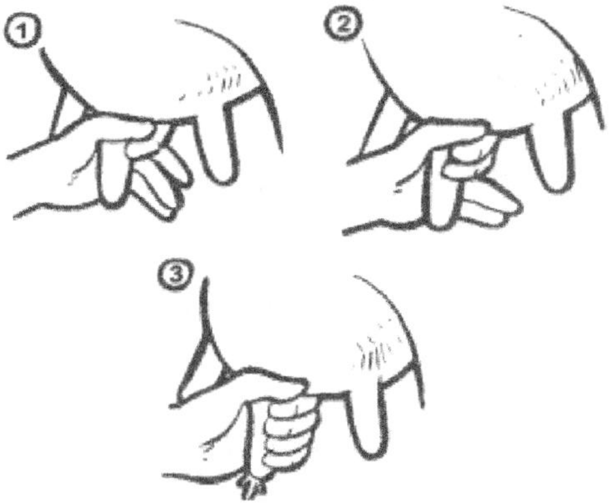

How to use Intra-mammary Antibiotics?

- Milk the infected teat completely (as shown).
- Clean the tip of teat with disinfectant.

- Insert the plastic canula (that comes with intra mammary antibiotic) into tip of teat.
- Attach the tube of antibiotic to the canula.
- Squeeze the contents of the entire tube into the teat.
- Remove the canula and hold the end of the teat closed.
- Gently massage the medicine up into the udder.

Caution

Some intra mammary antibiotics contain steroids that may cause abortion in pregnant animals. The label should be carefully read and followed.

Prevention & Control

- Keep animals clean, particularly after milking.
- Wash, clean and dry udder before milking.

Povidone Iodine: glycerine = 9:1

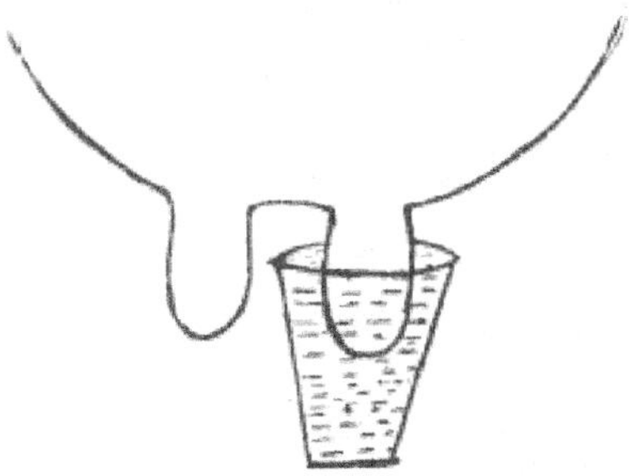

- Dip the teats in antiseptic solution (Povidone iodine) immediately after milking
- Take care not to injure teats during milking. The teat should be squeezed not pulled.
- Drying off (this is a period of time, where the animal is not being a milked in between lactations) animals should usually be dried off at the end of lactation.
- Soft bedding materials should be used. Concrete floors should be avoided.
- Wash hands of milkers before milking
- All the equipment and containers should be cleaned and washed properly.

Infertility

What is infertility?
When a female animal cannot regularly produce healthy offspring this is called infertility. There are many causes of infertility.

Causes of infertility:

1. Insufficient grass and feed
The female animal must develop eggs in her ovary before coming into heat. If the animal cannot get enough food, the eggs will not develop and she will not come into heat regularly.

2. Internal parasites
Worms and flukes eat the nutrients that the animal needs to reproduce. This will make the animal thin and weak and she will not come into heat.

3. Diseases of the reproductive organs

There may be disease of the reproductive organs (uterus, ovaries, cervix) caused by microorganisms. This may prevent the animal from coming into heat. If the animal is bred and becomes pregnant, she may abort the fetus. In some animals the reproductive organs may not develop properly. This will also cause infertility. The female, in female and male twins, in cattle are always infertile.

Signs of reproductive tract infection:

- Foul smelling pus and mucus may be passed from the vulva.
- Rectal examination of the uterus may show enlargement, lumps, or other abnormalities.
- If the animal has an infection called brucellosis, then it will abort in the last trimester of pregnancy. Brucellosis can be transmitted to humans, so be careful handling the afterbirth and fetuses.

4. Breeding at the wrong time

Pregnancy will result only if the male breeds the female at the proper time of her heat cycle. If the male breeds her too early or too late, pregnancy will not occur.

Prevention and treatment of infertility

1. Animals must have nutritious food.

- Give 3-4 kg of balanced feed daily for 2 weeks.
- Feed legume plants (beans, peas) on a regular basis.
- Allow the animal to lick salt block every day.
- Feed 1 CoFeCu® tablet daily for 20 days, and then give 3 Prajana® tablets daily for 2 days. If the animal still does not come into heat, give 3 more Prajana® tablets on days 11 and 12.

2. Give medication against internal parasites on a regular basis!

3. Treatment of reproductive tract disease:

Clean the uterus with a mixture of 1 part disinfectant (Dettol® or Savlon®) and 100 parts clean water. Povidone iodine solution can also be used in the uterus. It is best given on the day of heat. If there is a fever, antibiotic injections will be necessary.

4. Breed at the right time.

Cattle and buffalo should be bred within 12 to 18 hours of showing signs of heat. For sheep and goats, breeding should be done with 18 to 24 hours of showing signs of heat. Do not let the male mate one animal more than 3 times.

Breed Improvement

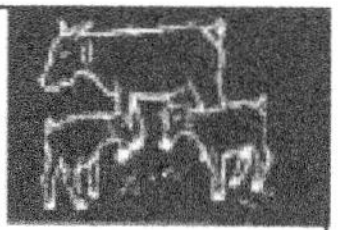

What is breed improvement?

The method of choosing good quality male and female animals. The goal is to improve production of meat, milk, and eggs and to produce improved offspring.

Qualities of a good male animal:

- Must be healthy and active.
- Choose a male whose mother has given birth to many healthy offspring.
- There should not be any problems with the reproductive organs.
- The body weight and height should be appropriate for the age and kind of animal.
- Choose animals whose parents have grown rapidly and given good production.
- They should be the best male animals available locally.
- Should have its own breed characteristics.

Qualities of a good female animal:

- Must be healthy and active.
- For goats, sheep, and pigs, choose a female that has given birth to large litters.
- Body weight and height should be appropriate for the kind of animal.
- Good udder development with a distinct milk vein in the cow and buffalo. Pigs should have 6-8 pairs of teats.
- High milk production.
- Healthy skin and smooth, shiny hair coats.
- Pelvis should be large to prevent birthing problems.
- Should have a history of regular heat cycles and good pregnancies.

Methods of heat detection:

Heat detection is determining the time of heat by observing the signs shown by the female. Animals show different kinds of signs and behaviours. The common signs and behaviours are as follows:

- Clear mucus is discharged from the vulva.
- The vulva is red and swollen.
- The animal will be restless and will vocalise (cry out) constantly.
- The animal will try to mount other animals and will stand quietly if a hand is placed on its back.
- Usually there is a decrease in appetite for a few days.
- There is usually a decrease in milk production 1 or 2 days before coming into heat.

Note: Signs of heat may be reduced when there is a calf suckling.

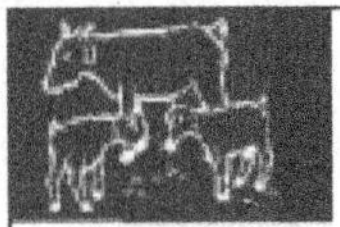

Right time to breed:

Different kinds of animals should be mated (bred) at different times. After the female shows signs of heat, breeding should be done at the following times:

- Cow, buffalo - 12 to 18 hours
- Sheep, goats - 18 to 24 hours
- Pigs - once at 12 hours, then once again 24 hours later

Methods of breed improvement

1. Avoid inbreeding

Breeding closely related animals is called *inbreeding.* For example: breeding siblings, father with daughter, or mother with son. This type of breeding is not suitable for breed improvement because:

- Breeding closely related animals will produce bad qualities.
- Ability to fight off disease will be less.
- Ability to produce healthy offspring will be less.
- Offspring may be unhealthy, weak, or deformed.

2. Practice outbreeding

Breeding non-related male and female animals is called *outbreeding.* For example: breeding local male and female animals that are of the same species, but do not share any common ancestors.

3. Breeding between different breeds (types) of animals

For example: breeding a local cow to an improved-type bull.

Advantages of breeding non-related animals:

- The offspring will tend to have the best qualities of both parents.
- They will have increased ability to fight off disease.
- This method will increase the production of more productive animals.

Digestive System Disorders

The parts of the digestive system are the mouth, esophagus (food pipe), stomach, intestines, rectum.

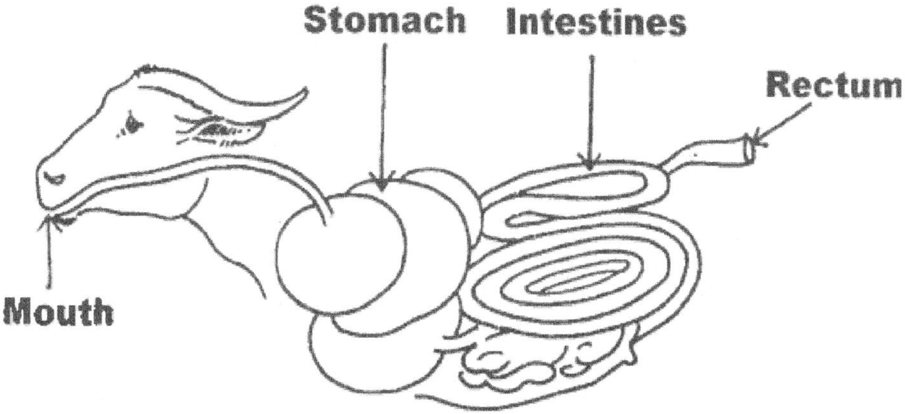

Some of the digestive system problems we can see:

1. Diarrhoea
2. Bloat
3. Loss of Appetite
4. Coccidiosis
5. Constipation

1. Diarrhoea

What is diarrhoea?

A frequent flow of thin, watery faeces is known as diarrhoea.

Which animals are affected by diarrhoea?

All domestic animals are affected.

What are the causes of diarrhoea?

- Infection with parasites.
- Infection with bacteria or viruses.
- Feeding rotten or mouldy feed.
- Eating large amounts of grain.
- Eating poisonous plants.
- Overeating of feed or water.

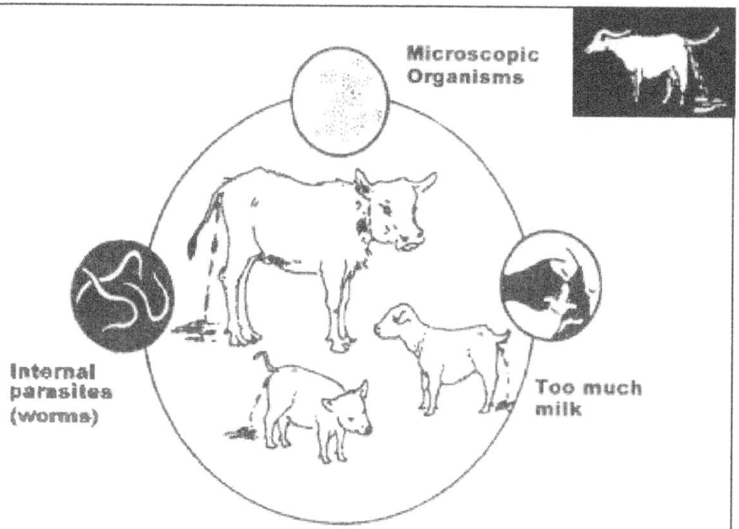

- An overdose of medications (especially C.T.C. for liver fluke).
- Overeating of young, succulent plants.
- Lack of vitamins.
- Medication given to treat constipation can also cause diarrhoea.

How do we recognise diarrhoea?

- Soft or watery faeces.
- Frequent defecation.
- Sunken eyes
- Thin or recent weight loss.
- Faeces may be stuck to animals skin and hair around the anus.
- Often increased intestinal sounds can be heard.
- The stomach may appear bloated.
- If the diarrhoea is due to microorganisms, the animal may have a fever.

Treatment:

Treatment must be given according to the cause

a) Diarrhoea due to parasites:

- Give Distodin®, Panacur®, Valbazen®, or piperazine as needed.

b) Diarrhoea due to the consumption of poisonous plants:

- Give sodium thiosulphate, magnesium sulphate, and Avil® injection.

c) Diarrhoea due to the overfeeding of grain or young plants:

- Give Neblon®.

d) Diarrhoea caused by microorganisms:

- Give sulphadimidine tablets.

Prevention:

- Regularly give medication to control internal parasites.
- Do not let the animals graze on areas with poisonous plants.
- Do not feed the animal too much grain or green feed at one time.

- Keep the shed, food, and water clean.
- Give nutritious food to the animals.

2. Bloat

This is a problem of ruminant animals. Gases produced in the stomach must exit through the mouth. If the gas cannot escape bloat will develop.

Causes of bloat:

- A sudden change in diet, particularly to more rich foods (such as green, leafy forage, grains)
- Overeating of grains.
- Overeating of legume plants, such as soybeans and other kinds of beans.
- Eating rotten or mouldy feed.
- Eating poisonous plants.
- Animals lying on one side for a long time.
- Lack of fibrous food in the ration.
- Choke (a condition when a small piece of wood, metal or any object gets stuck in the throat.

How do we recognise bloat?

- The Animal has difficult and noisy breathing.
- The animal may have it's tongue hanging out of it's mouth.
- Animals breathe with their mouth open.
- Stomach is enlarged. Tapping on the stomach produces a hollow, drum-like sound.

In the last stages, the animal lies on its side. If treatment is not given the animal will die in a few hours.

Treatment:

1. Open the mouth and place the animal with the head a bit higher than the body.
2. Give Timpol® and magnesium sulphate.
3. Give ample amounts of water.
4. If the above medications are not available, mustard oil or turpentine oil can be given.

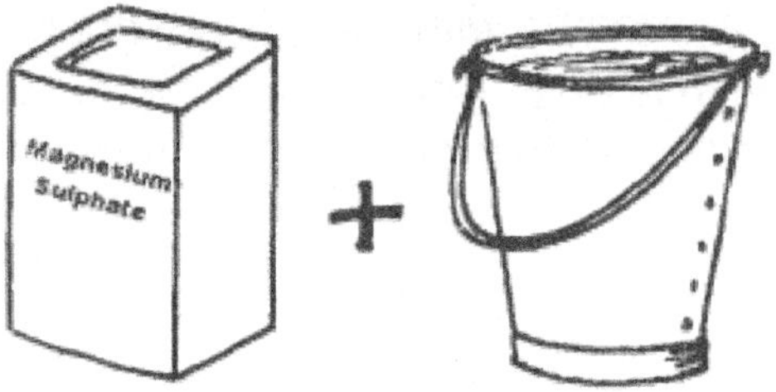

5. If possible, pass a rubber tube into the stomach to let the air escape.

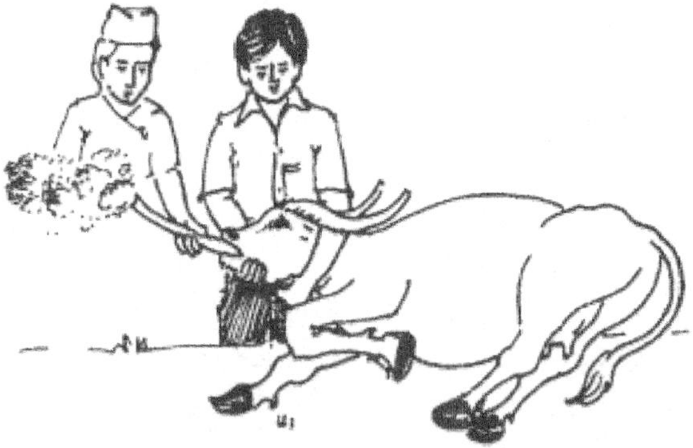

Last resort treatment:
If other treatments have not worked, prepare a clean area on the left flank (bloated side) of animal. Wash your hands and with a clean (flamed sterilized) knife puncture the side of the animal where there is the most tension. Take care nothing flammable is in the way of the gas being released. After making the puncture, apply ointments at punctured site to prevent infection & flies. Do not suture wound. Give antibiotic injections for 5 days. The wound may be kept open by using a clean (sterile) metal rod daily for next 5 days. This animal should be fed a high roughage diet.

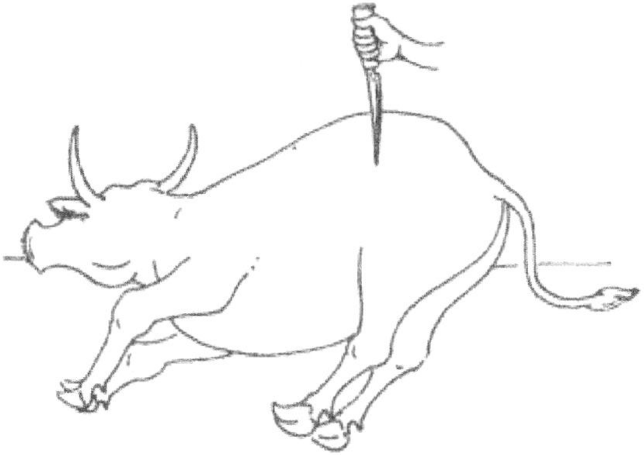

Home treatment:
Garlic, *siltimur,* grapefruit, cloves, and ginger have all been used to treat bloat.

Prevention:

- Avoid a sudden change in diet.
- Do not feed too much grain or legume plants at one time.
- Do not feed rotten or mouldy feed or grasses.
- If the animal is weak from an illness and recumbent (does not want to stand up) roll it from side to side.
- Do not let animals graze in areas where there are poisonous plants.
- Allow animals to drink before feeding them.

3. Inappetence

What is inappetence ?

When an animal is not eating, this is called inappetence. This problem is seen in all ages and all kinds of animals. There are many reasons why an animal stops eating.

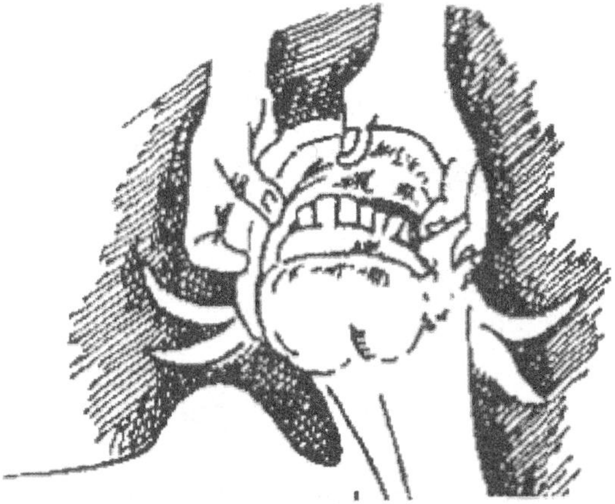

1. Wounds inside the mouth

There are many causes of wounds inside the mouth, such as:

- Foot and mouth disease or Orf (contagious exanthema).
- Punctures from thorns or wire.
- Burns from hot feed.
- Broken teeth.
- Swollen tongue.

2. Paralysis of the mouth

Animals with rabies or animals with tetanus are unable to swallow food. They cannot use the muscles in their jaw and tongue.

3. High Fever

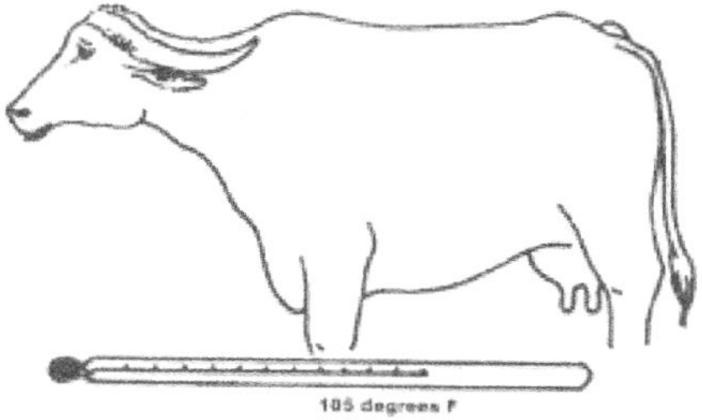

Animals with a high fever also do not eat, due to lack of appetite. As the temperature decreases its appetite is restored. There are many kinds of diseases that can cause a fever.

4. Poor quality food

- Poor quality food is not tasty. It may have a bad smell.
- Lack of green feed. The animal may get tired of eating dried grass all the time.
- Rotten and mouldy grass is not tasty.
- If the animals do not have sufficient water, they will not eat dry feed.

5. Worms or Liver Flukes

The stomach and intestines do not function properly if infested with parasites. Therefore the animals may not like to eat grasses and feed.

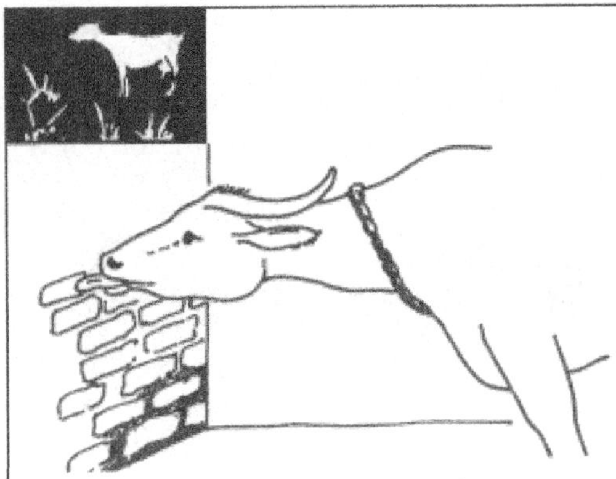

6. Mineral deficiencies

There are many types of minerals required by the body. Salt is one of these. If they are not sufficient in the diet, the animal may have no appetite. In this condition, the animal may lick sand, stone, brick walls, or other animals.

7. Overeating of grain

Sometimes an animal may get loose and overeat on grain because the grain is very tasty. The animal may then get a blockage and constipation and will not want to eat anymore food.

Methods of Treating Inappetence

The cause of the inappetence must be determined before treating.
We treat according to the cause. For example:

1. Wounds inside the mouth:
If the wounds become infected, then penicillin injections must be given to bring down the fever and kill the infection. These animals can be fed sugar or molasses mixed with water, or a mixture of flour and water. Salt, hot food, or rough grasses should not be given. The mouth should be washed out each day with a mixture of alum and water.

2. Paralysis of the mouth:
This could be due to rabies and tetanus. There is no treatment for rabies. Tetanus can be treated with high doses of penicillin or tetracycline.

3. High Fever:

- Find the cause of the fever and give penicillin or Terramycin® injections.
- Give plenty of water.
- Cover the animal with wet cloths.
- Keep the animal in an open, well-ventilated place.

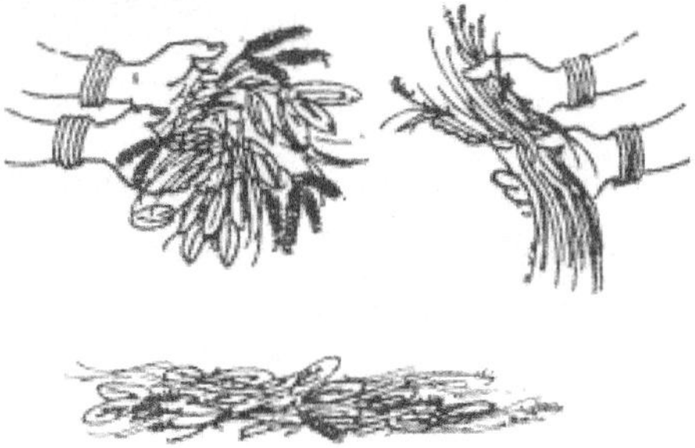

4. Poor quality grasses and feed

- Mix green grass with dry grass when feeding.
- During the dry season, give the animal more drinking water. Give water at least 3 times per day.
- When feeding straw, wet it or chop it into small pieces. Straw can be mixed with rice hulls.
- Do not feed rotten, moldy, or bad-smelling grasses.
- Give Battisa® medicine.

5. Worms or Liver Flukes

- Give the animals worm and liver fluke medication regularly.
- Give Battisa® medication.

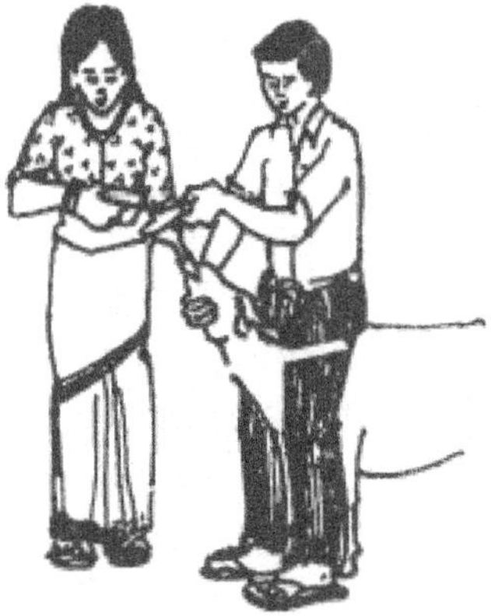

6. Mineral Deficiency

- 1 to 1½ soup ladles of salt should be given to a cow or buffalo each day.
- 1/4 soup ladle of salt should be given to sheep and goats.
- A homemade salt block can be made by mixing red soil, salt, and eggshells.
- Mix green grasses and dried grasses when feeding.

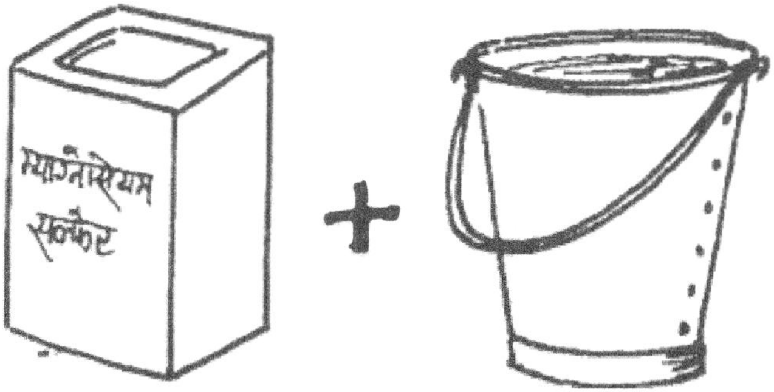

7. Overeating of grain

- Magnesium sulphate should be given
- Give lots of water
- Feed yeast

If none of the above reasons are the causes of inappetence and the animal appears healthy, then Battisa® should be given. A course of 2 or 3 days Vitamin B injections can also be given.

4. Coccidiosis

What is coccidiosis?

This disease is caused by microorganisms. Because it causes bloody diarrhoea it is sometimes called red diarrhoea.

Which animals are affected by coccidiosis?

All animals are affected, but poultry, nursing calves, kids, and lambs are most susceptible.

How do animals get coccidiosis?

- By eating contaminated dirt or faeces mixed with their food.
- If milk bottles or feeding tubes are dirty.
- If sheds are dirty and damp.

How do we recognise coccidiosis?

- Sudden onset of diarrhoea.
- Diarrhoea has a bad odor.
- Frequent, watery faeces.
- The faeces may have blood or foam in them.
- The animal may quickly become weak.
- The eyes may look sunken.
- The diarrhoea may be red in colour.
- The chickens may be dull and depressed and eat less feed.
- Post-mortem examination shows blood and water in the intestines.

Treatment:

Salt/Sugar solution (electrolyte)

Mix together:

- One pinch salt
- Half handful of sugar
- 3 glasses of water

Other treatment:

For Poultry

- Amprolium is also recommended. It is mixed with the water and fed to the chickens.

For Young Animals

- Give sulphadimidine tablets (Dose in medicine book).
- Electrolytes as above

Prevention:

- Feed the animals sufficient colostrum immediately after birth.
- Keep the animals and chickens in a clean, dry, well-ventilated shed.
- Change the bedding regularly.
- Feed orphan calves, kids, and lambs with clean utensils.
- Amprolium powder can be mixed with the water to prevent the disease.

5. Constipation

What is Constipation?

It is the condition when the animal have unusually hard stool. The cause may be eating dry food without enough water, or not eating enough roughage.

Symptoms:

The animal has no appetite does not chew its cud, has no stool, may appear uncomfortable, and may strain to defecate, but without success.

Treatment:

Ordinary cooking oil, or mineral oil, is a very effective treatment. For a big animal, 1-4 liters of oil should be fed with a stomach tube or bottle. If the oil is too expensive, then try giving magnesium sulfate. Provide several buckets of water. Since the animal may not want to drink, give the water by stomach tube, bottle, or bamboo. Offer fresh, green grass to help stimulate the digestive tract.

Control / Prevention:

Animals should always be given lots of fresh water to drink. Fresh, green grass or other roughages are also necessary to keep the stomach moving properly.

Respiratory Diseases

There are two general respiratory conditions:

a) Pneumonia

b) Cough

a) Pneumonia

What is pneumonia?

- It is caused by microscopic organisms.
- These organisms grow most rapidly if the shed is damp, humid, and poorly ventilated.
- When the animal breathes, these organisms travel to the lungs. They attack the lungs and cause disease.

Which animals are affected by pneumonia?

All animals are affected.

How do we recognise pneumonia?

- Fever.
- Difficult and rapid breathing.
- Watery discharge from nose that later becomes thick and white.

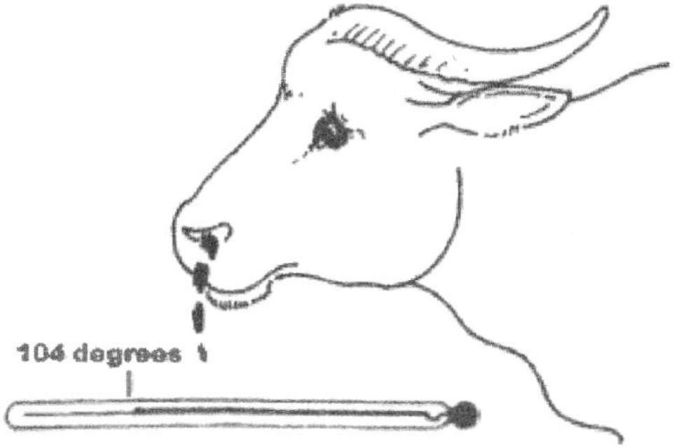

Treatment:

- Procaine penicillin
- Penidur® (benzathine penicillin)
- Terramycin® or Terramycin L.A®

Prevention:

- Keep animals in a warm, dry, and well-ventilated shed.
- If the animal gets wet during a cold winter rain, dry it thoroughly with a towel.
- Change the bedding material regularly.

- Keep sick animals away from the healthy ones.
- Do not give the sick animal's feed to the healthy ones.

b) Cough

What is a cough?

This is actually not a disease. It is a common sign of respiratory diseases such as small roundworm infection, pneumonia, and airway problems. Many diseases can cause coughing. Sometimes food and water can become stuck in the windpipe and cause coughing. All animals can get a cough.

How do we recognise a cough?

- Coughing can be heard even when the animal is at rest.
- Often the conjuctiva (skin around eyeball) is red.

Treatment:

- It is not good to give liquid medicine to a coughing animal.
- Make small cubes of Caflon® or Cafgo ® mixed with sugar and feed these to the animal.
- Give Avil® injection.
- If the cough is caused by parasites or pneumonia, then treatment for those diseases must be started.

Prevention:

- Control parasites.
- Make sure the shed is well-ventilated and dry.
- Be careful when drenching animals so as to not injure the throat.

Haemorraghic Septicaemia
(HS, Pasteureliosis)

What is it?

- This is a disease caused by microscopic organisms (a *microscopic organism* means that they cannot be seen with the eye alone).
- It is a contagious disease with a very sudden onset.
- It is known by various names in Nepal such as: *deshaan, hawa, ghyaar ghyaar, ghokre,* and *24-hour disease.*

Which animals are affected by H.S.?

Cattle, buffaloes, sheep, goats, and pigs.

How do we recognise H.S.?

- Rapid development of signs.
- The throat area is swollen, hard, and hot to the touch. Sometimes the entire head may be swollen.
- There is a high fever (104-107°F).
- The animal has difficulty breathing.
- A raspy sound comes from the throat (*ghyaar...ghyaar* sound) while breathing.
- Breathing is often done with the mouth open.

- The animal drools saliva.
- Slight bloating of the stomach.
- Discharge from the eyes & nose can be seen.
- The animal can die quickly, often in a matter of hours.

Treatment:

- Tetracycline or Sulpha drugs can be used.

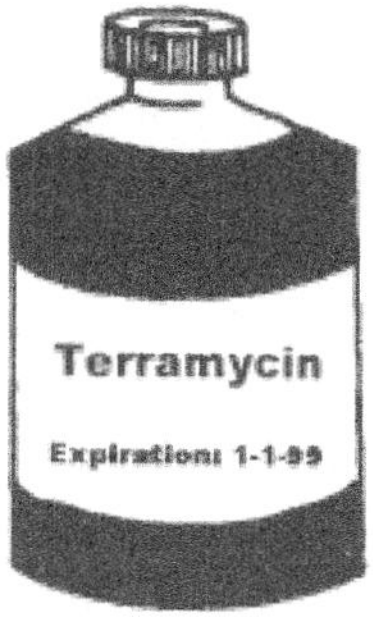

Prevention:

- Keep the sick animal away from the healthy animals.
- Do not use the feed and utensils used by the sick animal for feeding healthy animals.
- Take care of the healthy animals first, then treat the sick ones.
- Bury the dead animals and burn the bedding material.
- Give H.S. vaccine every 6 months.
- Report the outbreak to the concerned authority immediately.

Black Quarter

What is it?
This is an acute infectious disease caused by microscopic organisms. It has a very sudden onset.

Which animals are affected?
Cattle, buffaloes, sheep, and goats. Mostly it is seen in fat, healthy animals of 8 to 24 months of age.

How do we recognise black quarter?

- Incubation period of disease is 2-5 days.
- Animal has a high fever (106-108°F).
- The thigh, shoulder, or neck of the animal becomes swollen.
- The swollen parts are hot to the touch and give a crackling sound when rubbed.
- If treatment is not given, the animal will die.
- Lameness is usually one of the first symptoms.
- Rapid breathing.
- Animal may die within 12 to 48 hrs. after onset of symptoms.
- Grunting may be heard due to pain.

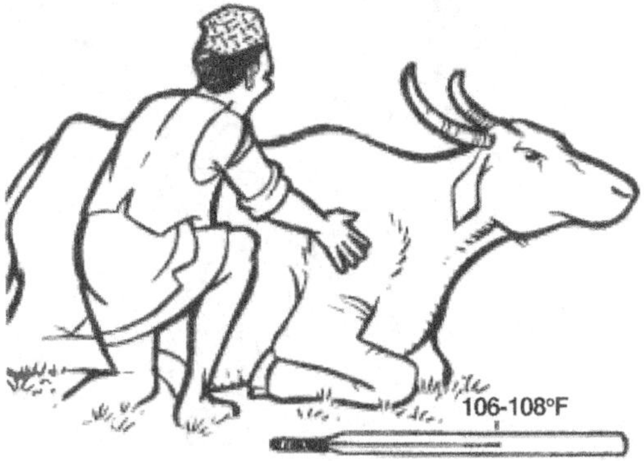

Treatment:

Give 2 doses of procaine penicillin injection to the animal for the 1st day, then give 1 dose each on the second and third days. Treatment can be continued for 5 days. If penicillin is not available, Terramycin® can be used in the same way.

Note: In acute cases animals may die even with treatment.

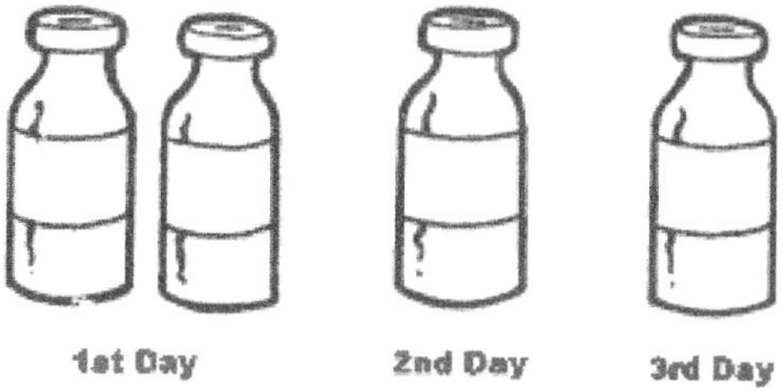

Prevention:

- Do not mix sick animals with healthy ones.
- Bury the dead animals. If possible a layer of lime should be put directly on the body before it is covered with soil or stones.
- Give black quarter vaccine every 6 months.
- Report the outbreak to the concerned authority immediately.

Foot and Mouth Disease (FMD)

What is FMD?

This is a contagious disease that is caused by a virus. The disease affects the mouth and the feet. Nursing calves and imported (western) breeds of cattle may die from FMD.

Which animals are affected by FMD?

- All ages of cloven hoof (split hoof) animals are affected. Horses are not affected by this disease.
- Cattle & buffalo are more susceptible. Mortality (death) is found more in young calves.

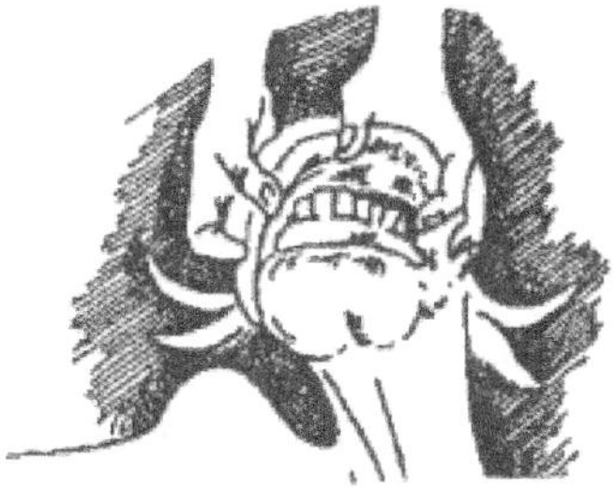

How do we recognise FMD?

- Incubation period of this disease is about 2-3 weeks.
- First sign is fever (104-106°F).
- Animals do not eat.
- Animals drool saliva.
- Wounds or ulcers may be seen on the gums and the tongue.
- Abrupt reduction of milk yield.

- Wounds may be seen between the toes *(interdigital area)* and the coronary band.
- Sometimes wounds may be seen on the teats and udder.

- Often there is lameness.

- Imported (western) breeds of cattle may die from FMD.
- Pregnant animals may abort.

Treatment:

- There is no specific treatment for FMD, but symptomatic treatment can be done.
- Clean the foot wounds and apply ointment.
- If the wounds are infested with maggots, the wounds can be cleaned and then treated with turpentine oil to kill the maggots.

- Wounds inside the mouth can be cleaned with alum solution. If there is lots of salivation, then cleaning the mouth wounds is difficult.
- Feed affected animals soft, cooked food like porridge.
- Feed nursing calves milk using a bottle or bamboo pipe.
- The animals should be kept in a clean dry shed, not in a muddy place.

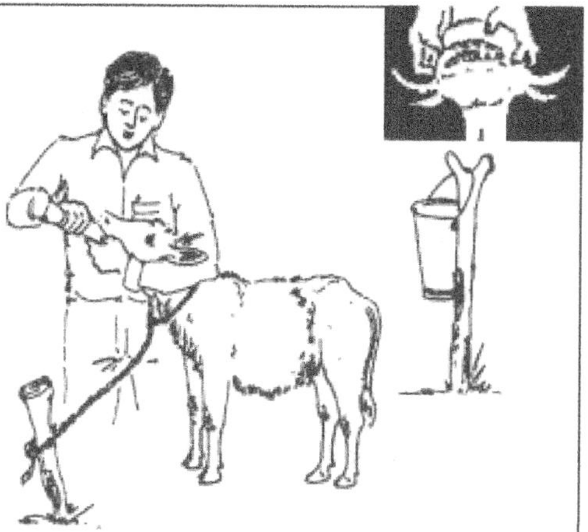

Prevention:

- Vaccinate animals against FMD every 6 months.
- Keep sick animals away from healthy ones.
- Do not mix utensils and feed used for sick animals with those used for healthy ones.
- Take care of healthy animals first, then sick ones.
- Feed the animals in their stall. Many types of animals may come and graze in the same pasture. Stall-fed animals will not be exposed to sick animals, so the risk of spreading the disease is less.
- Report the outbreak to the concerned authority immediately.

PPR (Peste Des Petitis Ruminants) / Goat Plague

What is PPR?

PPR is an acute and a highly contagious disease of small ruminants. Usually this disease spreads in hot summers. This disease is also called "Pseudorinderpest"

Cause

PPR is caused by virus called Morbillivirus. This PPR virus is closely related to the virus causing Rinderpest.

Which animals are affected?

The disease affects sheep and goats only. Cattle buffaloes, pigs and other animals are not affected by this disease. Goats are more susceptible than sheep.

Mode of Transmission

- Direct contact with sick animals
- Inhalation (contaminated air)
- Contaminated food, water, bedding and other appliances
- Secretions (discharge from nose, eyes) and excretions (faeces)
- This disease is not transmitted through insect bites.

How do we recognize PPR?

- High fever (104-105°F) and a dry muzzle.
- Watery discharge from eyes and nose at first, and later becomes thick like pus which gives a bad odour (smell) and sneezing.
- Develop pneumonia-like symptoms and coughing.
- Sores/erosive lesions (wounds) seen in mouth (gums) that look like wheat grain.
- Fever may last for 5-8 days.
- Severe diarrhoea which leads to dehydration occurs. Diarrhoea may contain blood and mucus.
- Post-mortem and lab examination should be done for proper diagnosis.
- Death usually occurs 7-12 days after the onset of fever.

Treatment

- There is no specific treatment for PPR, but symptomatic treatment can be given for diarrhoea, and respiratory problems.
- Broad spectrum antibiotics can be given to stop secondary bacterial infections. This may decrease the number of animals that die.
- Oral fluids can be given to reduce diarrhoea and dehydration.

Prevention / Control

- All animals should be vaccinated at the age of 3-4 months.
- In the case of an outbreak, ring vaccination should be done (vaccinate in a circle around the affected geographical area, but not in it).
- Sick animals should be kept separated from other animals. Strict sanitation and hygienic measures need to be adopted.
- Disinfection of food and water utensils, bedding materials used by sick animals should be burned.
- Report the outbreak to the concerned authority immediately.

Swine Fever (Hog Cholera)

What is Swine fever?

It is a highly contagious viral disease of pigs. It has a rapid onset and many pigs are affected at the same time.

Which animals are affected?

Pigs of all ages are affected by this disease.

Mode of Transmission

- Direct contact with infected pigs
- Disease may enter through eating waste food or contaminated feed and water
- Urine, discharge from nose and eyes

How do we recognize Swine Fever?

- High fever (106-108°F)
- Vomiting
- Discharge from nose and eyes
- Purplish / purplish-red patches found on abdomen, inner side of legs, and ears

- Tremor, convulsions
- Become weak and wobbly in hind legs
- Most pigs die within a week, although a few may recover

Diagnosis
- History taking
- Post mortem
- Laboratory test

Treatment
There is no effective treatment for swine fever. To stop secondary bacterial infections antibiotics should be used.

Prevention and Control
- Vaccinate piglets at the age of 42 days.
- Sick pigs should be isolated
- Waste food should be properly cooked before giving to pigs
- Food residues and bedding materials used by sick animals should be burned, and utensils should be disinfected.
- Report the outbreak to the concerned authority immediately.

Foot Rot

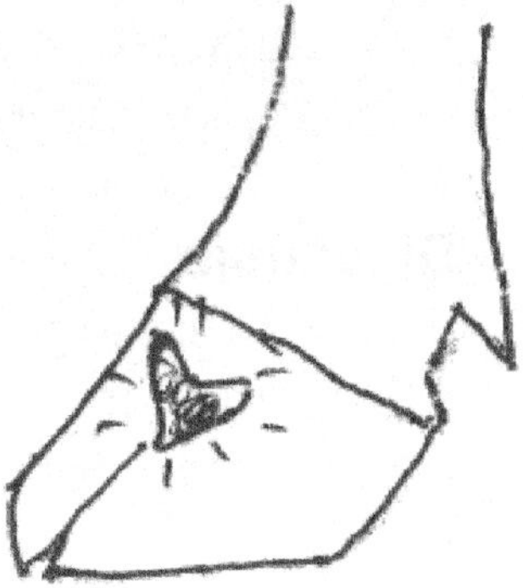

What is foot rot

Foot rot is a highly infectious disease affecting the feet of sheep and goats. The infection starts between the toes, then the hoof starts to rot until the sides of the hoof separate from the bottom of the hoof. Foot rot is commonly seen in the rainy season.

Cause

It is caused by two bacteria working together; *Fusiformis necrophorus* and *Dichelobacter nododus*. Generally these organisms inhabit mud and soil. They gain entry to the animals' feet when the animal is kept in a wet, muddy yard for a long time, or when the foot gets injured or wounded.

The environmental factors predisposing to infectious foot problems are wet, muddy yards or poorly drained pastures, overgrown hooves, overcrowding, and introduction of infected goats or sheep to a susceptible flock.

Which animals are affected?

All cloven-footed animals are affected by this disease, but sheep and goats are most commonly affected. Cattle and buffalo are occasionally affected. Clinical findings in goats are less severe than in sheep.

Mode of Transmission

- Transmission occurs by contact between infected and susceptible animals at pasture under suitable conditions of moisture and temperature.
- The organism is present in discharges from animals with clinical lesions and can invade the damaged epidermis of the foot of susceptible animals.
- Carrier state may also occur in animals.

How do we recognize Foot Rot?

- Severe lameness, such that the animal may not even want to stand up.
- Poor condition and weight loss since the animal does not eat properly.
- Red, swollen tissue around the coronary band and between the toes
- Sores between the toes that may smell badly and ooze pus.
- Hoof wall separating or splitting.

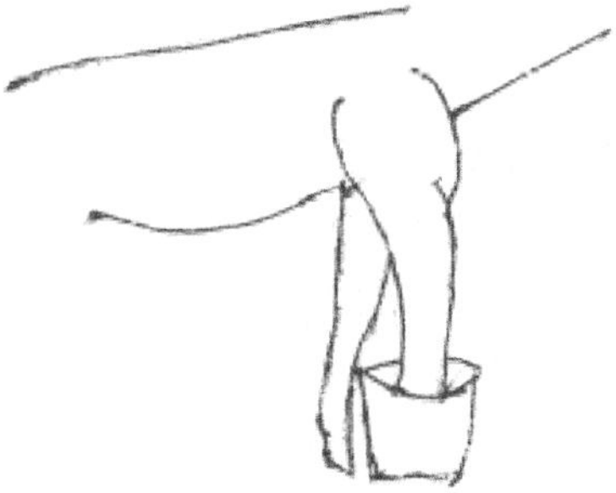

Treatment

- Trim the hoof well and clean the wounds with antiseptic solution such as povidone iodine. All the pockets of pus must be opened, and dead tissue removed.
- Apply antiseptic ointment and bandage the feet.

Or

- Disinfectant footbaths can be used with various disinfectant chemicals, such as: 10 % Copper Sulphate, 10 % Zinc Sulphate or 5-10 % formalin. Animals' feet should be clean before they use a foot bath, they should be at least one minute in a foot bath.

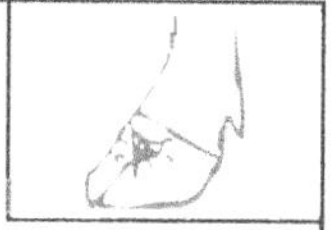

- If formalin is used, the solution should be changed daily to insure that evaporation has not resulted in excessive concentration.
- Antibiotic injections (broad spectrum) will help control severe infections
- keep the animal in a clean, dry environment
- Change the bandage whenever it becomes wet or dirty.

Prevention/Control

- The animals can be walked through a footbath of 10% formalin
- Keep the animal in a clean, dry environment
- Trim the hooves regularly
- Do not buy animals from herds that are known to be infected.

IBD (Gumboro)

What is IBD?

This is a very infectious and contagious disease of poultry. It spreads very quickly among young chickens. Chickens 6-18 weeks may become sick, but it is quite rare. Young chickens of 3-6 weeks of age are mostly affected by this disease. Chickens lose their ability to resist other diseases because the bursa fabricius is affected in this disease.

Cause

IBD Virus

Mode of transmission

- Direct contact
- Contaminated feed and water
- Litter, faeces
- Clothes, insects, mosquito

How do we recognize IBD?

- A number of chickens get sick at the same time.
- Whitish, watery diarrhoea is common, along with a soiled vent
- Ruffled feathers
- Chickens become uncoordinated
- Loss of appetite
- Dull and depressed

During post mortem, the following changes can be seen

- Bursa Fabricius is swollen and may have red spot in it.
- Pin point haemorrhage (bloody spot) may be seen in between proventriculus and gizzard
- Chest muscles and leg muscles are often swollen and may have blood spots in them
- White granules may also be seen in kidney.

Note: Report the outbreak to the concerned authority immediately.

Prevention of IBD

- Cleaning and disinfection of houses between flock and practice of "all in all out" system.
- Regular vaccination against IBD.
- Bury the dead chickens.
- Keep sick chickens separate from the healthy ones.
- Do not mix utensils and feed used for sick chickens with those used for healthy ones.

Newcastle Disease (Ranikhet)

What is Newcastle Disease?

- This is a contagious disease of poultry caused by a microscopic organism (a virus).
- Sometimes 100% (all) of the chickens in a flock will die of this disease. Other names for this disease are *rudha* and *qhumaune.*
- Sometimes 3% of layers may survive.

How do we recognise Newcastle Disease?

- Chickens do not eat feed but will drink.
- The wings drop down and the birds drag them on the ground.
- The head and neck become twisted.
- The affected birds turn round and round in the same place.
- Paralysis is seen in the legs and wings.
- There is a watery discharge from the nose and mouth.
- There is a sudden decrease in egg production.
- The birds have difficulty breathing and make lots of respiratory noise.
- The chickens will die within a few days despite treatment.

Treatment:

None.

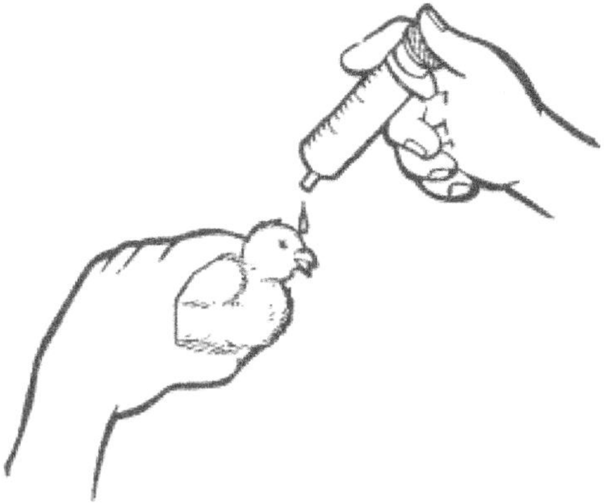

Prevention:

- Vaccinate against Newcastle disease at the beginning of an outbreak.
- Bury the dead chickens.
- Keep the sick chickens separate from the healthy ones.
- Do not mix utensils and feed used for sick chickens with those used for healthy ones.
- Clean the chicken pen and spray phenol solution (Phenyle®) to kill the virus. Keep the pen empty for 3 weeks, after which time new birds can be placed in the pen.
- Report the outbreak to the concerned authority immediately.

Fowl Pox

What is Fowl Pox?

This is a contagious disease caused by a virus. Once affected the chickens will slowly die.

How do we recognise Fowl Pox?

- Sores and blisters may be seen on the featherless parts of the body.
- The blisters will burst and a scab will develop.
- There may be sores in the mouth, on the tongue and throat so the chickens cannot eat properly. Breathing may also be difficult.
- There is nasal discharge.
- Tears and scabs around the eyes make it difficult for the chicken to see. If they cannot see they will not be able to find their food.
- There may be so many sores around the eyes that the eyeball is destroyed.

Treatment:

There is no treatment for fowl pox, but some supportive care may help.

- Clean the sores and apply Himax® ointment. Treat the sores daily until they go away.
- Himax® should not be used in the eyes. Use alum solution only to clean around the eyes.
- If the chickens are force-fed by hand, the death rate may be less.

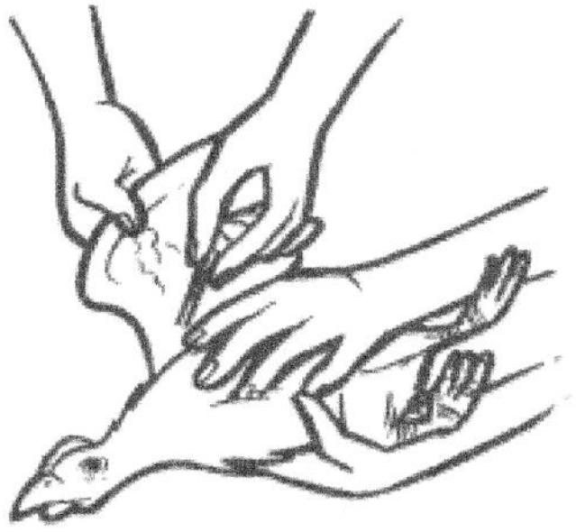

Prevention:

- Vaccinate against fowl pox at 6-8 weeks of age. In the case of layers vaccinate at the age of 12-24 weeks.
- Burn or bury the dead chickens.
- Keep the sick chickens separate from the healthy ones.
- Change the bedding regularly and keep the chicken pen clean.

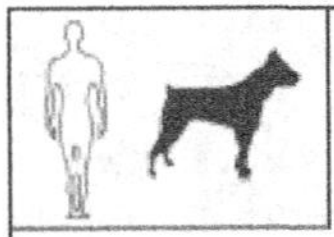

Zoonotic Diseases

Introduction

What are Zoonotic Diseases?

Diseases that are transmitted from animal to human, or human to animal are called Zoonotic diseases.

Some Zoonotic Diseases are:

1. Brucellosis
2. Anthrax
3. Bird Flu
4. Trichinosis
5. Cysticercosis and Taeniasis
6. Rabies

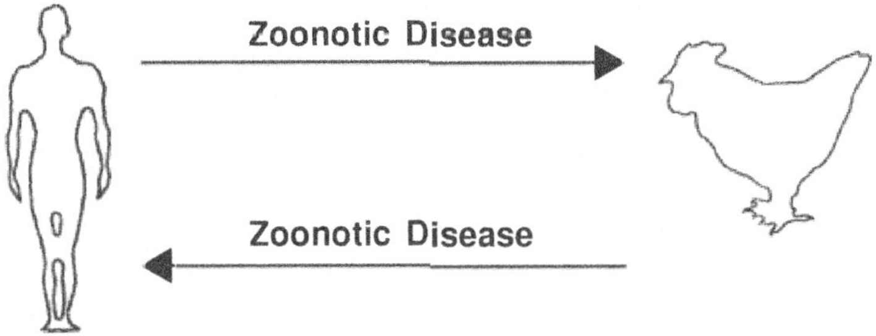

Mode of Transmission

1. Air droplets, urine, faeces

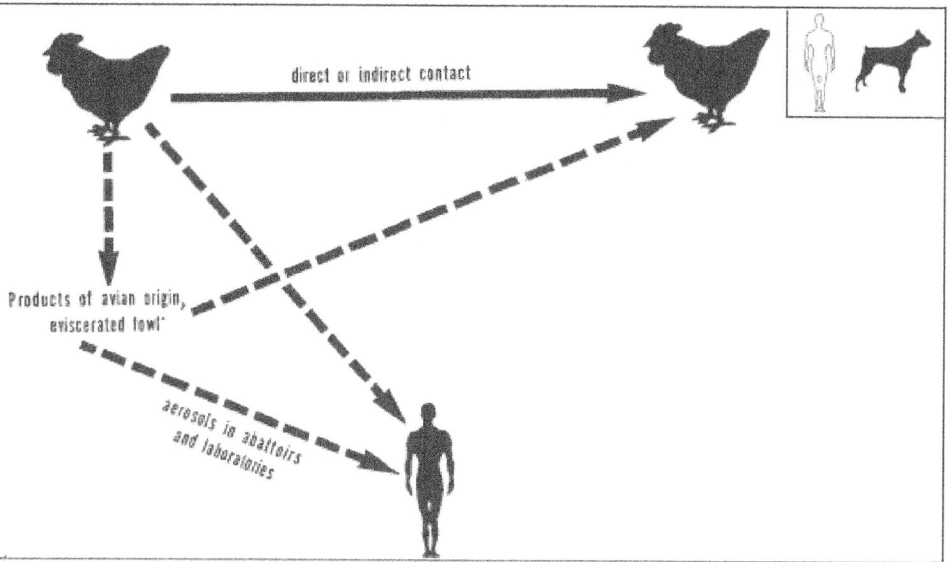

2. Animal products like milk, meat, skin, wool

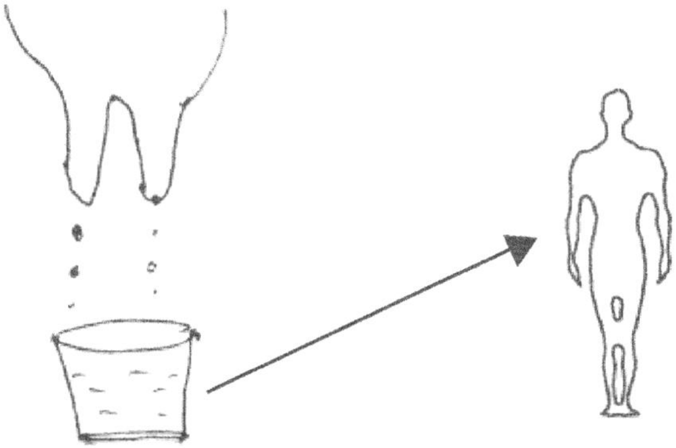

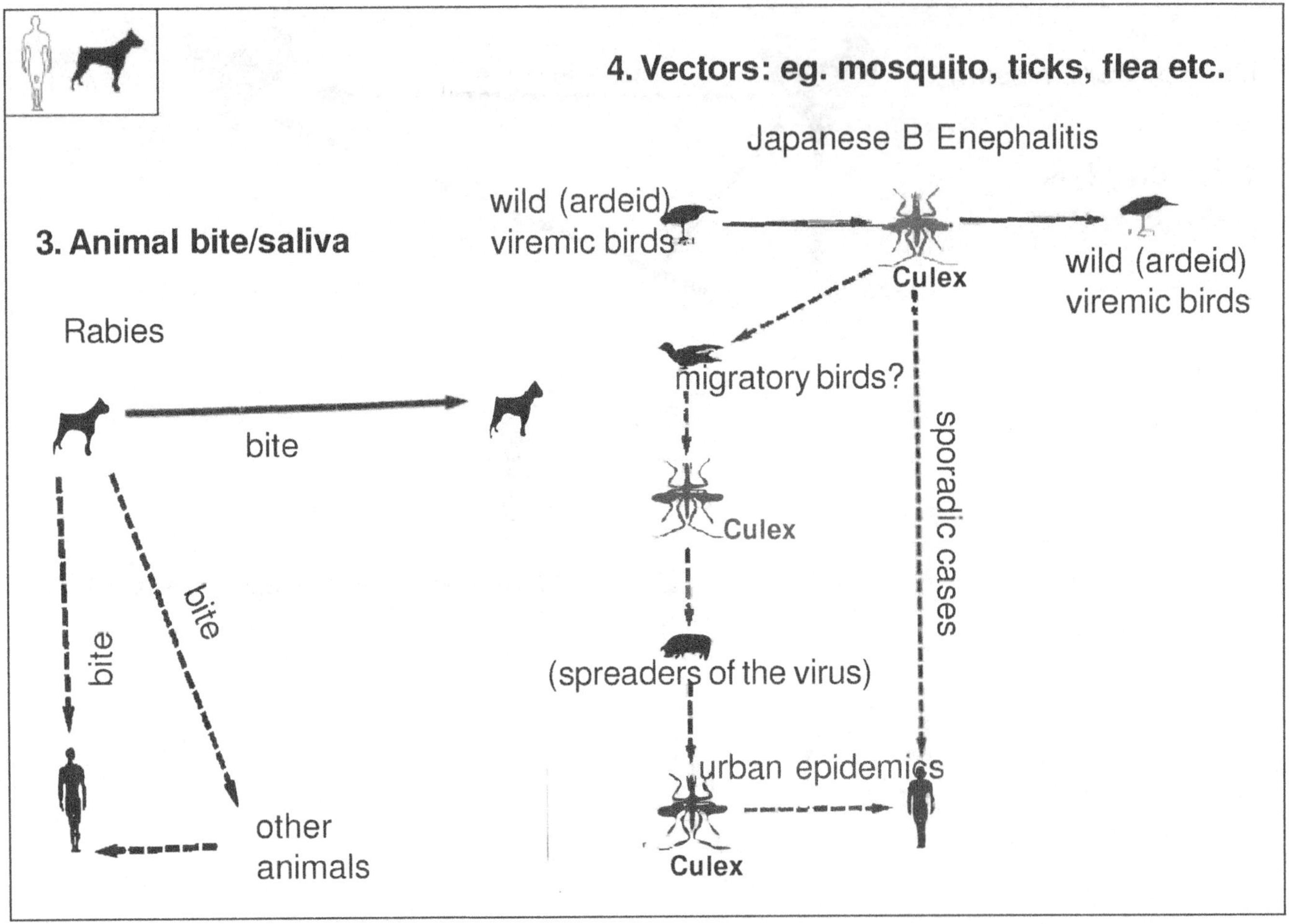
4. Vectors: eg. mosquito, ticks, flea etc.
Japanese B Enephalitis
wild (ardeid) viremic birds
Culex
wild (ardeid) viremic birds
migratory birds?
Culex
(spreaders of the virus)
sporadic cases
urban epidemics
Culex
3. Animal bite/saliva
Rabies
bite
bite
bite
other animals

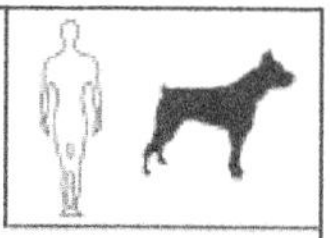

Some of the Common Zoonotic Diseases

1. Brucellosis

What is it?

It is a contagious bacterial disease of domestic animals that causes abortion

How does it transfer from animal to animal or man?

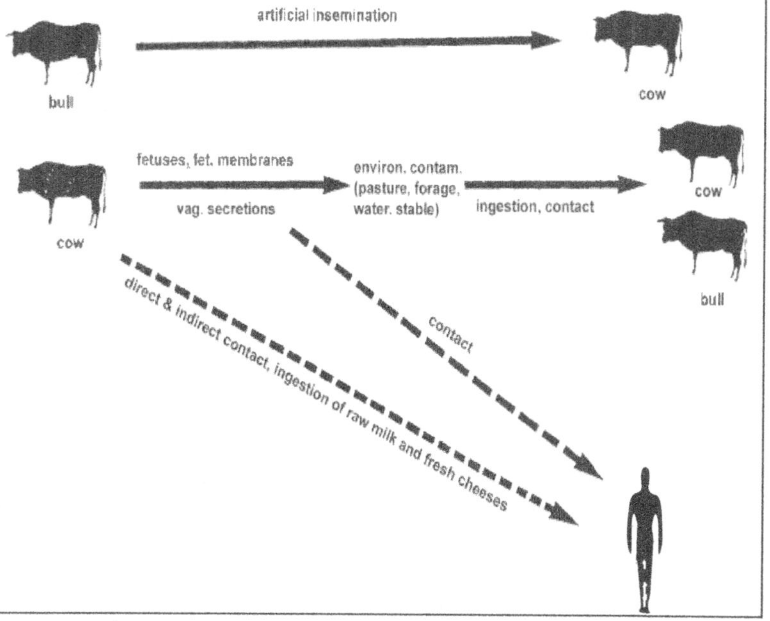

1. By the ingestion of feed or water contaminated with aborted fetus, placenta, vaginal discharge or blood.
2. The bacteria may enter the body through broken skin, or conjunctiva.
3. During coitus of infected bull or cow.
4. Human may get infected by ingestion of contaminated unpasteurized milk or meat or direct contact with infected material (i.e. aborted fetus etc.).
5. Human to human transmission is rare.

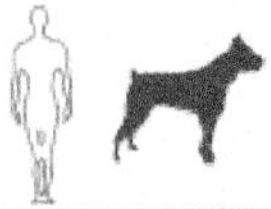

How to recognize brucellosis in:

Animals

1. In cattle it causes abortion during the 6th and 9th month of pregnancy, often with retained placenta. Mastitis may occur.

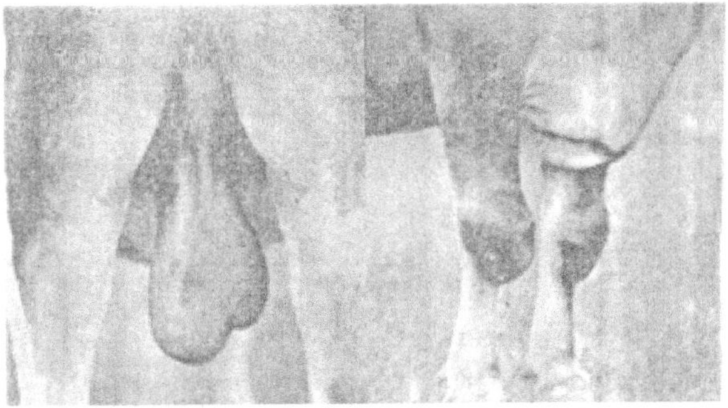

2. In bulls, the symptoms are orchitis (inflammation of testis), arthritis and swelling of the knees.
3. A subsequent infection may render the animal infertile.

In humans

1. High intermittent or wave-like fever (therefore called an undulant fever)
2. Arthritis and orchitis
3. Weakness, body pain and sweating
4. Headache, lack of appetite

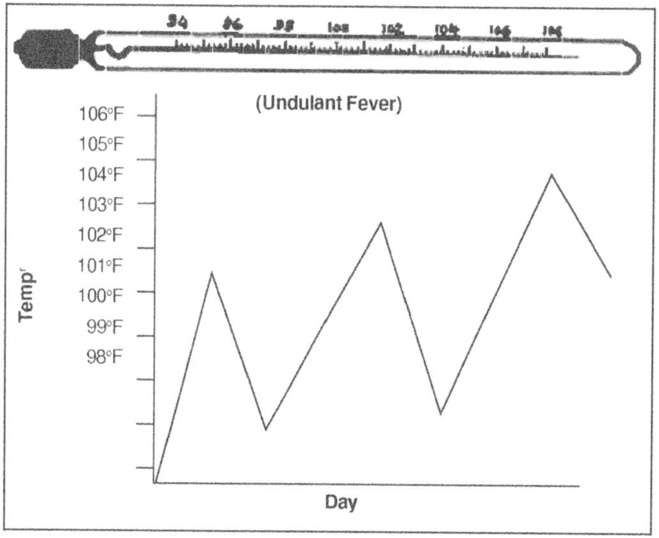

Treatment and Control

Treatment of Brucellosis in animals is not satisfactory.

For control

1. Isolate the suspected animal.
2. If the animal is Brucellosis positive, it is recommended to dispose of the animal.
3. The aborted fetus, placenta and vaginal discharge should be disposed properly away from water sources and human dwellings.
4. Report animal outbreak to concerned authority immediately.

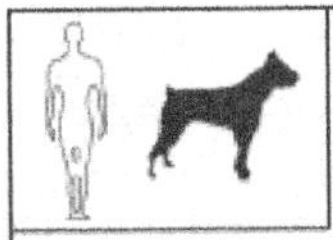

2. Anthrax

What is it?

Anthrax is an acute, contagious bacterial disease of mammals especially of herbivores. It is an occupational hazard of people affecting agriculture, animal husbandry and industry who are in frequent contact with animals and their products.

Mode of transmission

1. Bacterial spores play an important role in transmission of the disease.
2. Contact with blood and discharge from the infected animal.
3. Contaminated feed, water, and animal by-products.
4. Meat, milk, skin, and wool.
5. Mechanical carriers like birds, insects and flies.
6. Animal health workers (like meat handlers), leather and wool processors (therefore called "wool sorters" disease), and other animal husbandry workers are at high risk of infection.

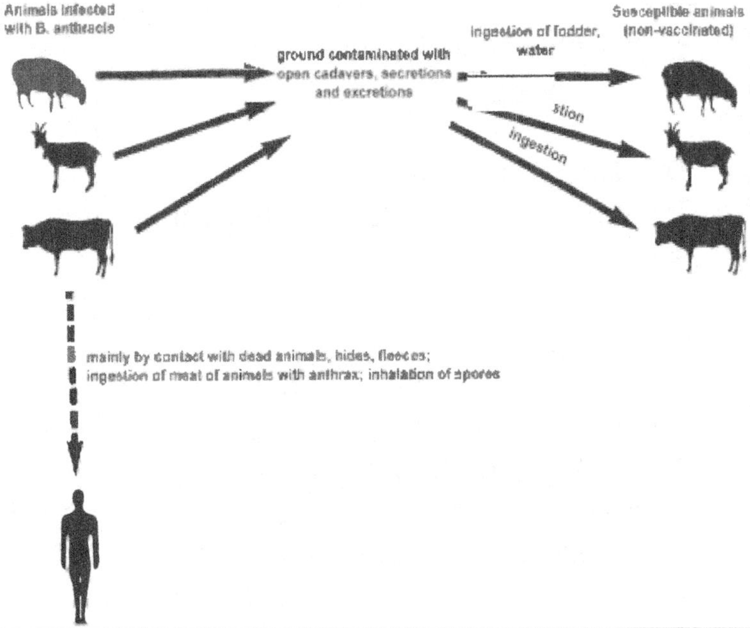

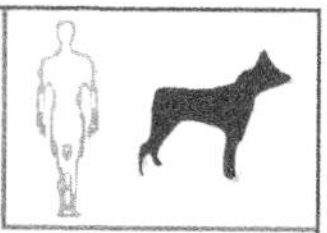

How can we recognise an anthrax animal?

1. The animal may die suddenly without showing any clinical symptoms.
2. In an acute case the animal shows a high fever, digestive disturbances (like bloat), muscle tremor, and difficulty in breathing. The animal may die within 2 days.
3. Following death there is oozing of dark-coloured blood from all the natural orifices (mouth, nostrils, ear, rectum, vagina, penis), which fails to coagulate.
4. The carcass (dead body) fails to develop rigor mortis.

What are the symptoms in man?

Anthrax occurs in humans in three forms: cutaneous or skin, inhalation, and gastro-intestinal anthrax.

Cutaneous anthrax

Symptoms: presence of pustules on the skin

Inhalation anthrax

Lungs get infected after inhalation of the spores.
Symptoms: fever, difficulty in breathing, headache, and chest pain

Gastro-intestinal anthrax

It occurs after ingestion of spores contained in foods; especially meat, milk, and milk products.
Symptoms: stomach pain, fever, vomiting, and the presence of blood in faeces
Death occurs due to shock.

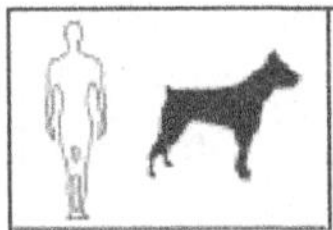

How can we treat anthrax animal?

If noticed in the early stages, give high doses of penicillin or tetracycline intramuscularly for 3-5 days.

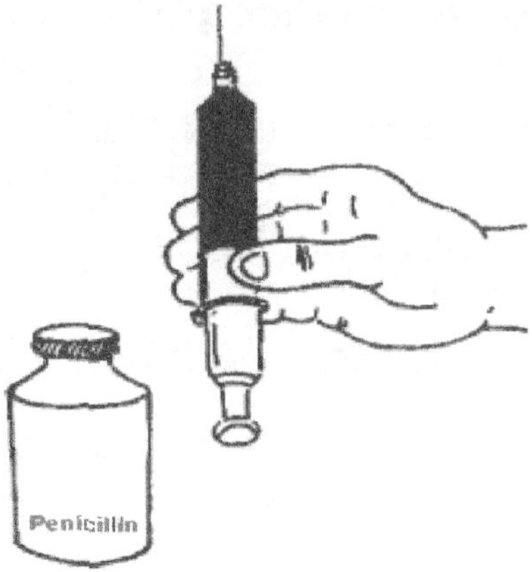

Prevention and control

1. Dispose of carcass by burning or by deep burying with sufficient lime (calcium carbonate) to cover.
2. Disinfect the premises and farm appliances with 10% formalin solution. Vaccinate animals against anthrax in endemic area.
3. Vaccination can be done at any age of animal.
4. Report animal outbreak to concerned authority immediately.

Note: If you suspect that the animal has died of anthrax, NEVER do a post mortem examination!

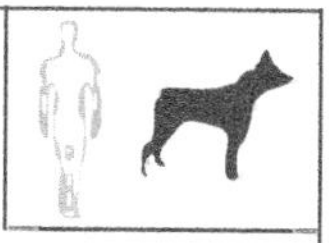

3. Bird Flu (Avian Influenza)

What is it?

It is a serious infectious and contagious disease of birds - especially of poultry, duck, turkey and wild fowls. It is caused by a virus that leads to mass death of flock within 48-72 hours.

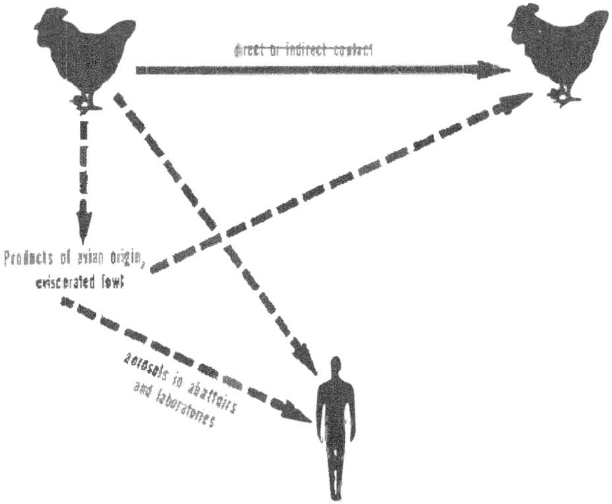

How does it transfer from bird to bird or man?

1. Birds which are in direct contact with infected birds or virus contaminated food, water, utensils, vehicles, clothes, etc.
2. Humans get infected by the ingestion of raw or undercooked meat and eggs.
3. Bird flu is an occupational hazard of people who are in frequent contact with poultry and poultry products eg: poultry farmers, poultry transporters, meat processors, laboratory workers, and veterinarians.

How to recognise bird flu?

In birds

1. Birds of all ages die suddenly without showing any clinical symptoms.
2. The head may be swollen, the colour of comb and wattle change to pale.
3. There is a mucous discharge from the nostrils.
4. Loss of egg production in adult females.
5. Difficulty in breathing
6. Greenish diarrhoea
7. The bird may show neurological disorders like paralysis of the neck.
8. Post mortem examination shows bleeding in trachea, proventriculus and joints.

In humans

High fever, muscle pain, conjunctivitis, headache, cough, infection of throat, stomach ache, and diarrhoea are common symptoms. Humans may die within 3 to 30 days of infection. To confirm the disease, it needs laboratory examination.

What are the preventive measures?

1. Infected and dead birds should be disposed of properly.
2. Maintain strict Bio security on the farm.
3. Reporting to concerned authorities should be done promptly if the disease is suspected.
4. Eggs and meat should be cooked to 150^0 F (70^0 C) for human consumption.

4. Trichinosis

What is it?

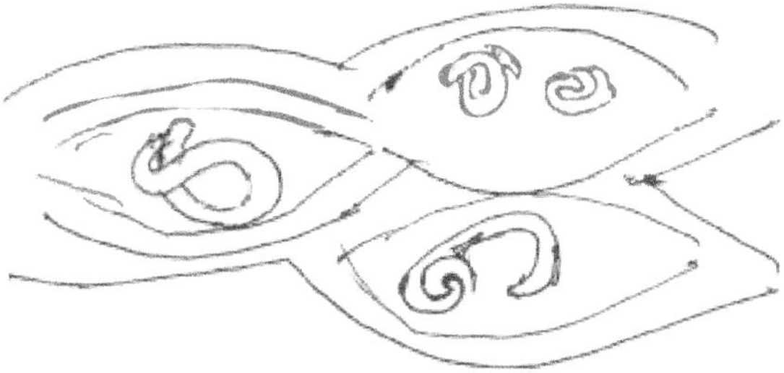

Trichinella in the muscle of a pig

It is an infection caused by small roundworms called *Trichinella spiralis* that lodge in the small intestine of pigs. This worm is also found in rats and bears.

How do humans get infected?

Humans get infected by ingestion of larva present in the muscles of pig meat which is not well cooked.

How does it transfer from pig to man?

The female worm in the intestine of a pig penetrates the intestine wall and lay eggs which hatch into larvae. These larvae move to various muscles of the body, like the diaphragm, tongue, cheek muscles and other muscle tissues. After a month the larvae coil up to form a cyst. Other animals or people get infected by eating this cyst in the muscles. The larvae in the cyst develop into adults in the intestine of the new host, and produce more larvae in various muscle tissues.

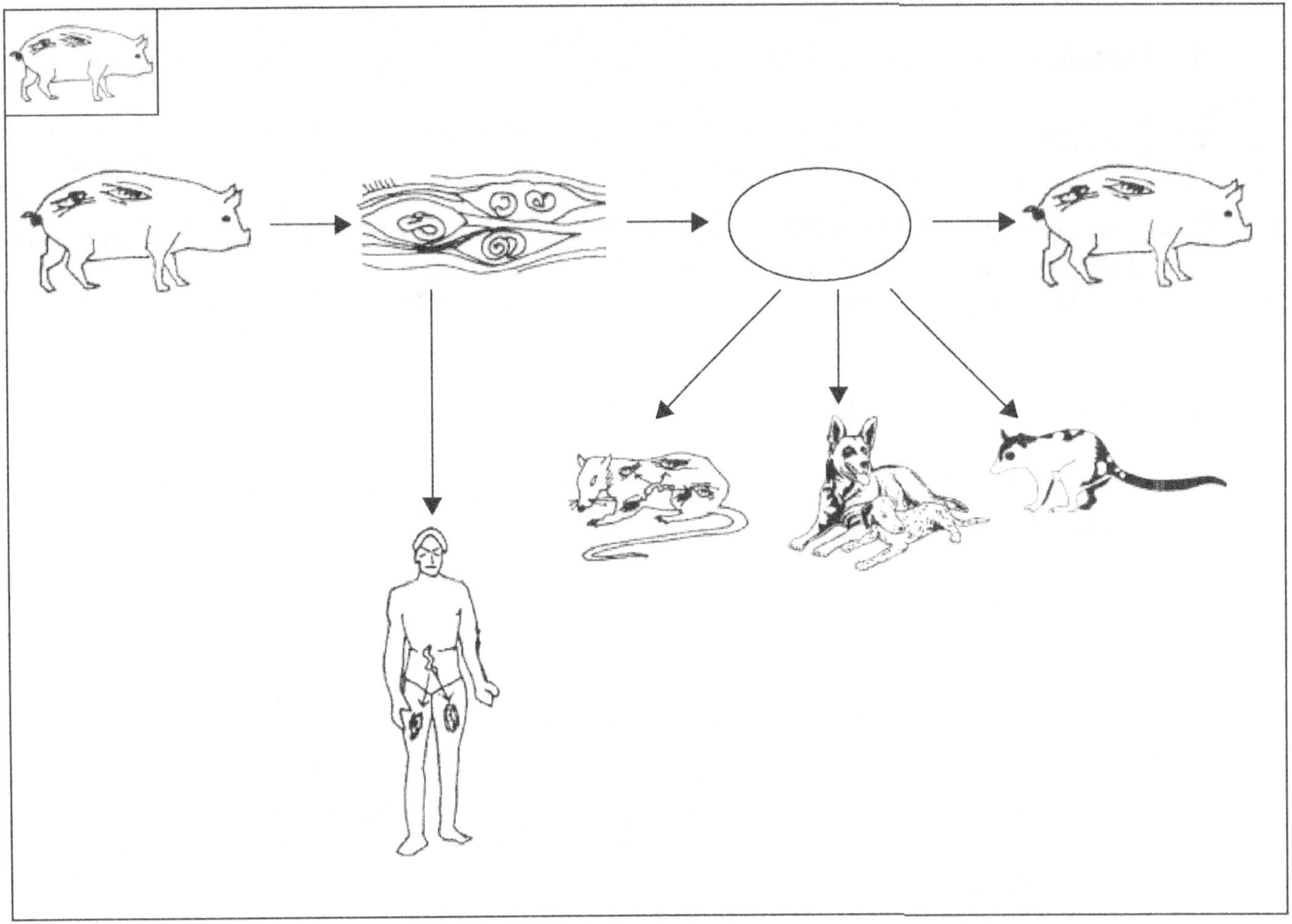

What are the symptoms of Trichinosis?

In animals

Pigs can tolerate trichinosis well, so they do not show symptoms unless they are heavily infected. The symptoms are muscle pain and weakness, incoordination of limbs, and death.

In man

Initially there is nausea and vomiting, diarrhoea, abdominal pain, and weakness for a few days. Later, there may be respiratory disorders like difficulty in breathing, muscular pain and weakness, fever, visual problems, and nervous disorders.

Treatment

Treatment in pigs is impractical.

Prevention and control

1. Proper disposal of waste and offals.
2. Do not allow the animals to gain access to undercooked garbage and offals.

3. Pork should be cooked properly at more than 140^{0}F (60^{0}C) to kill the cyst present in the muscles.
4. Pork can be stored at 0^{0}F (-18^{0}C) for at least 24 hours to kill the cysts.

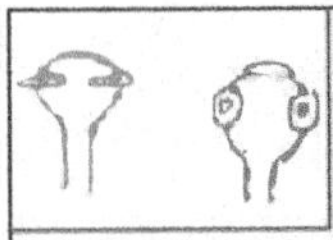

5. Cysticercosis and Taeniasis

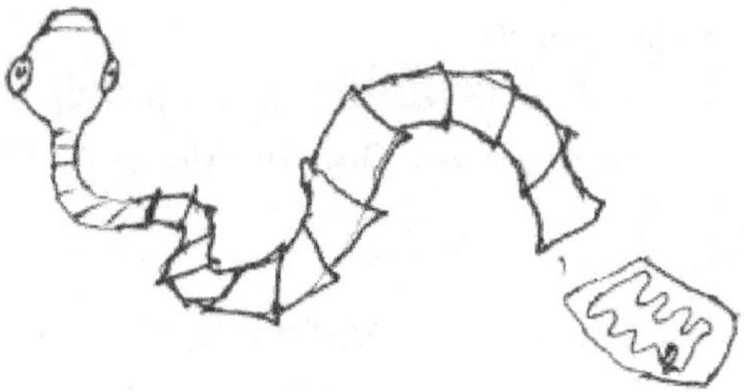

What is it?

It is an infection caused by tapeworms, namely *Taenia solium* and *Taenia saginata*. An infection caused by its larval stage is called Cysticercosis, and infection caused by adult tapeworms is called Taeniasis. The adult tapeworm grows to a length of 3-8 meters in the intestine of humans.

Which animals are susceptible?

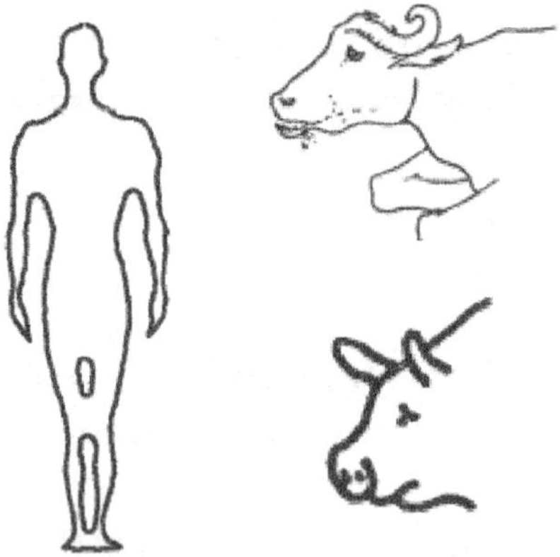

To complete the life cycles of this tapeworm, it needs two types of hosts.

a) Final/ Definitive host eg: Man

b) Intermediate host eg: Man, Cattle, Pig

How does man get infected?

Man gets infected by eating undercooked beef or pork.

Life Cycle:

Eggs secreted by humans are ingested by pigs or cattle, and they hatch in the intestine. Embryos penetrate the intestine and localize in the liver via the blood stream. Then they are distributed to various parts of the body like the heart, diaphragm, tongue and other muscular tissues where they become cysts. They also localize in the brain, kidney, and lungs. Man gets infected by ingestion of the insufficiently cooked meat which contains the cysts. The cyst in the human intestine develops into an adult tapeworm. Sometimes people accidentally ingest the eggs secreted by the tapeworm. The cyst develops in different parts of the body as in pigs or cattle. A cyst in the brain is called *Neurocysticercosis.*

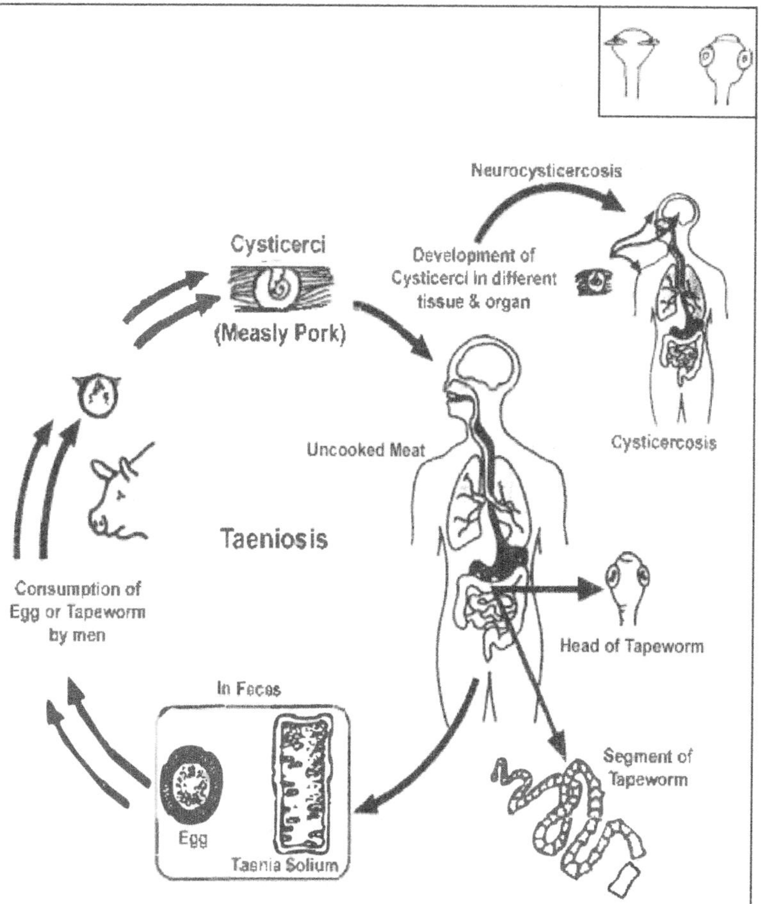

What are the symptoms

In pigs or cattle

The cyst can remain alive for many years in cattle or pigs. There are no apparent symptoms in the host. Most cysticercosis cases are detected during post mortem examination.

In man

The adult tapeworm infection in man does not create serious problems. However larval infections, where the cyst may be present in the brain, heart, below the skin, or in the eye may cause serious effects on the body.

The common symptoms of cystercosis are:
Epilepsy, headache, weakness, loss of weight, abdominal pain, nausea, vomiting, and eye problems. Confirmation of the disease needs x-ray and other laboratory examinations.

Control and prevention

1. Every effort should be made to break the life cycle of the tapeworm.
2. Animals should not be allowed to come in contact with human faeces.
3. The cyst in the animal tissues should be properly disposed of by burying.
4. Proper meat inspection should be made to avoid human consumption of cysts.
5. Storing meat at -4°F (-20°C) for 12 hours will kill the cysticerci present in meat.
6. Always eat sufficiently cooked meat.

6. Rabies

What is rabies?

This disease is caused by a microscopic organism (a virus) and is spread through the bite of a *rabid* animal (an animal with rabies). It is also known as *mad dog disease.*

Which animals are affected by rabies?

Mainly animals that live in the jungle, such as jackals, jungle cats, and monkeys. Among domestic animals, dogs are mainly affected. Any warm-blooded animal can get rabies, including man.

How do we recognise rabies?

- After getting rabies, the sick animal will change its behavior.
- For example: dogs may become angry, bite anything, bark continuously, or run away from the house.
- Cattle and buffaloes may become afraid of people, break their ropes, and run continuously. They develop a hoarse voice.
- Affected animals lose their appetite and drool saliva.
- They may attack anything they see such as people, trees, or other animals.
- The rear parts of the body may become paralysed within 7 days of the onset of the disease.
- The sick animal will die within 14 days of when the signs first appeared.

Treatment:

- There is no treatment for the disease once signs have appeared!
- Kill and bury the animal in a deep pit. If the rabid animal has bitten or scratched any other animals or people, treat those wounds by cleaning thoroughly with soap and water.
- Clean the wound 3-4 times and get the advice of a veterinarian. Give your animal rabies vaccine if recommended.

Prevention:

- Animals that have died of rabies must be buried in a deep pit.
- Tie up the rabid dog as this prevents other animals and people from being bitten.
- Vaccinate your own dog against rabies every 12 months, starting from 3 months of age.
- Control and remove street dogs.
- Report animal outbreak to concerned authority immediately.

Hygiene and Disease Prevention

Hygiene (cleanliness) is an important part of animal husbandry. Raising animals without cleanliness will lead to the spread of disease and death of our animals. If we give clean grass, feed, and water to our animals and keep them in a clean shed we can prevent disease from the very beginning.

What is disease prevention?
It is a plan or a method to prevent disease from occurring. There are many methods we can use to prevent animal disease.

Disease prevention methods:

1. Feed colostrum

Colostrum (first milk) increases the capacity of the young animal to fight off infectious diseases. Feed plenty of colostrum to a newborn animal.

2. Give nutritious feed

Animals of all ages should be fed plenty of nutritious grass and feed. This makes the animals strong and healthy, and enables them to fight off diseases. It also allows the pregnant animal to give birth to healthy offspring.

3. Build a well-ventilated shed that is easy to keep clean.

A dirty shed or barn is a source of disease. Many different diseases come from a dirty shed, such as: skin diseases, respiratory diseases, worms, fleas, and lice.

In addition to cleaning the shed regularly, chemicals (disinfectants) can be used to keep the disease organisms under control:

- Spread lime powder on the floor of the shed regularly.
- Mix 1 part potash with 100 parts water and spray in the shed.
- Mix 1 part phenol (Phenyle®) with 100 parts water and spray in the shed.

4. Stall-feeding:

Diseases can be spread from communal pastures and watering holes. Sick and healthy animals meet at these places. A disease can then spread from a sick animal to a healthy one. This problem does not occur when animals are fed and watered in their own stall.

5. Keep animals clean

- Lice, fleas, ticks, and mange can affect animals that are kept in dirty, dark conditions.
- Animals should be given a bath from time to time. Use a rough brush or comb to clean the hair.

6. Regular deworming

Livestock that receive medication for worms and liver fluke on a regular basis will stay healthy.

7. Vaccination

Vaccines for different infectious diseases are available. Disease organisms are not able to grow in vaccinated animals. Vaccines must be given *before* the animal becomes sick. Some vaccines that are available: HS, FMD, Black Quarter, and Rabies.

8. Do not mix healthy animals with sick animals.

Many diseases can be spread from sick animals to healthy ones. They can be spread by the animal handler and by the utensils used for feeding and watering.

- Separate sick animals from the rest of the herd.
- Use separate utensils for sick and healthy animals.
- First take care of the healthy animals, then treat the sick ones.
- Burn the bedding of sick animals.

9. Dead animals should not be left lying about.

Bury the dead animal in a deep pit. Spread lime powder in the pit. This kills the disease-causing organisms. If the dead animals are left lying out, disease may spread.

Importance of Water

Water can be obtained from many different sources, such as the tap, pond, river, or streams.

When do animals need extra water?

- When suffering from diarrhoea.
- When they spend lots of time grazing in the sun.
- When they are working (plowing or pulling a cart).
- When they have only dry grass and straw to eat.
- When the animal is on any medication.
- During the hot season.

During the above conditions, animals of all ages should be given plenty of water!

Why is water needed by the animal?

- To produce meat, milk, and eggs.
- To control body temperature so it does not get too high (hot).
- To make blood.
- To make the body function properly.
- To digest food.

- To clean the body of waste products (urine, faeces, eye and nose discharge).
- To spread nutrients around the body in the blood stream.
- To protect from, and help in the treatment of kidney diseases.
- To make the joints flexible and strong.
- To excrete poisonous substances from the body.

Water from the tap or mountain stream is the best for animals. They should not be given water that is muddy or green.

How to give water?

- Always keep clean water in clean utensils near the animal.
- Feed plenty of green grass because it also contains lots of water.
- If the animal will not drink because it is sick, mix water with flour, rice bran, and salt and encourage the animal to drink this mixture.

Balanced Feed

What is a balanced feed?

Different types of food items are needed to keep animals healthy, to help them grow properly, and to achieve good production. Feed that is prepared with the right amounts of these necessary materials is called a balanced feed.

Why should we give a balanced feed?

- To increase the production of eggs, milk, and meat.
- To build bones, flesh, and hair coat.
- To help wounds heal quickly.
- To make the animal strong so it can work well.
- To protect the animal from disease.
- To help the fetus grow strong inside its mother.
- To help young ones grow rapidly.

How do we make a balanced feed?

A balanced feed can be made easily at home. Collect the following food items as shown below:

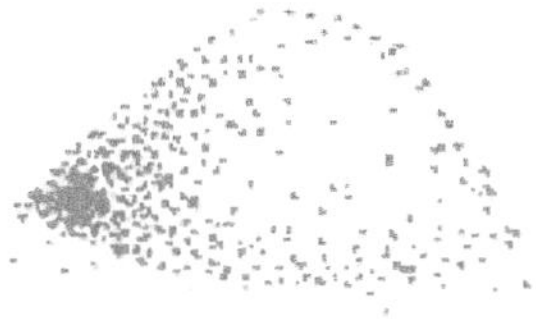

Rice bran - 2 parts

Maize, wheat, or millet flour – 1 part

Mustard cake or legumes (beans) – 1 part

All these items should be mixed well and fed to the animal.

How much do we feed?

If the animals are given lots of balanced feed, they will grow fast and stay healthy. It is not good to give this balanced feed by itself. Grass, hay, and straw should also be given together with the balanced feed.

For cattle and buffaloes give 4 to 6 mannas daily (about 2-3 liters)

For sheep and goats give $\frac{1}{2}$ to 1 manna daily (about $\frac{1}{4}$ to $\frac{1}{2}$ liter)

For pigs give 1 to 3 mannas daily (about $\frac{1}{2}$ to 1-$\frac{1}{2}$ liters)

The amount of feed depends on the condition of the animal, its age, body weight, and production. Pregnant and nursing animals should be given more feed than other animals.

Things to remember:

- Mustard cake and soybeans must not be fed raw. They should be cooked first and made into a powder.
- The prepared feed must be stored in a clean, dry, airtight container.
- Do not use the feed if it becomes mouldy or smells bad.

How to harvest fodder and forage

What is fodder?

Trees and bushes whose leaves can be used for animal feed are called *fodder*. Usually these trees and bushes grow near the farms or are planted on the terraces. Bringing this fodder from the forest to the home has been widely practiced in Nepal.

Leaves as well as young branches have been used as animal feed. Some of these plants are called: *tanki, kutmero, dudhilo, kanyu, nemaro, chuletro, katus, khimbu, and gogan.*

How and when to harvest?

- It is a good idea to use fodder during winter season, when there is a scarcity of green grass. Fodder leaves will fulfill the animals' requirements for green grass.
- It is best if a fodder tree is cut only once per year, although it is possible to cut 2 to 3 times per year from 1 tree.
- The best time to harvest leaves from a tree is when it is flowering. The leaves are most nutritious at this time.
- Fodder bushes should be cut before the leaves begin to fall on their own.
- Fodder leaves should be cut from the small branches. If large branches are cut, the next growth will be less.
- It is best to leave the fodder in the sun for a few hours after harvesting. This procedure is called *wilting*. It makes the fodder safer by removing any poisonous chemicals that might be in the plant.

What is forage?

Grasses and young plants that are raised around the farm for animal feed are called forage. These grasses usually grow by themselves, or they can be cultivated. Some common plants used as forage in Nepal are *banso,* maize, *berseem, dinanath, sentro,* signal, and napier.

How and when to harvest?

- These grasses can be planted during the monsoon and winter. They can be cut 2 or 3 times per year.
- If cutting is done only once per year, it is better to cut when the plants are starting to flower. The grasses will be most nutritious at this time.

- If the grasses are to be harvested 2 or 3 times per year, leave about 20 cm of the stalk. This will make it possible to get another cutting in about 1 month.
- It is a good idea to leave the harvested grass in the sun to wilt it. This will help remove any poisonous chemicals in the plants.
- If the production of forage is sufficient, then the grass can be dried and fed to the animals when other feed is not available. This type of dried grass is called *hay*.

Things to remember:

Young, tender plants should not be fed to animals. Some grasses may be poisonous at this stage. For example: *siris, gayo, angeri,* and *jwar* grasses.

Methods of feeding straw

Cattle and buffaloes eat many kinds of grasses and fodder, both fresh and dried. Ruminants prefer green feed to dried feed. Compared to green feed, straw is dry, hard, and not tasty. The dried straw can be chopped and soaked to improve its quality. This will help to improve the production of the cow or buffalo.

How do we improve the quality of straw?

- Estimate the amount of straw to be eaten by the animal.
- Collect this amount of straw.
- Using a sickle or knife, cut the straw into small pieces about 3-5 cm long.

- Place the straw pieces in a bucket.
- Add enough water to cover the straw. Salt can be added to improve the taste.
- Soak the straw for 24 hours.
- Soaked straw can be fed by itself, or mixed with other feed.

What are the benefits of improving straw quality?

- The chopped and soaked straw is more nutritious than dry straw.
- The chopped and soaked straw is tastier, so the animals eat more of it.
- The animals are healthier and stronger and give more milk.
- It is a good use of local resources.

Castration

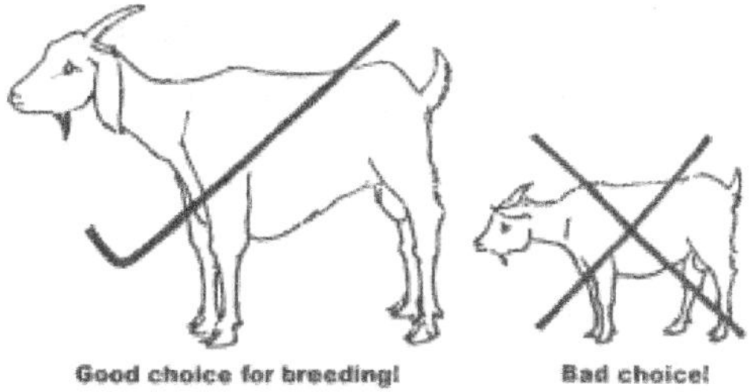

What is castration?

This is the method used to make a male animal unable to produce offspring.

Why do we castrate animals?

- To stop inbreeding.
- To prevent breeding of poor quality animals.
- To improve meat quality and taste.
- To make bulls more docile and easier to handle.
- To improve the breed.

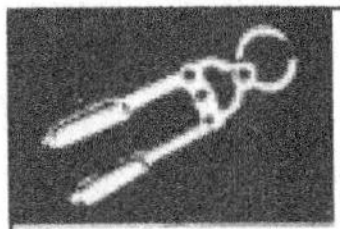

How do we castrate an animal?

There are many methods that have been used. The easiest way is to use an instrument called a *Burdizzo*. Male cattle, buffalo, sheep, and goats can all be castrated by using this method. The Burdizzo will not work on the pig or the horse.

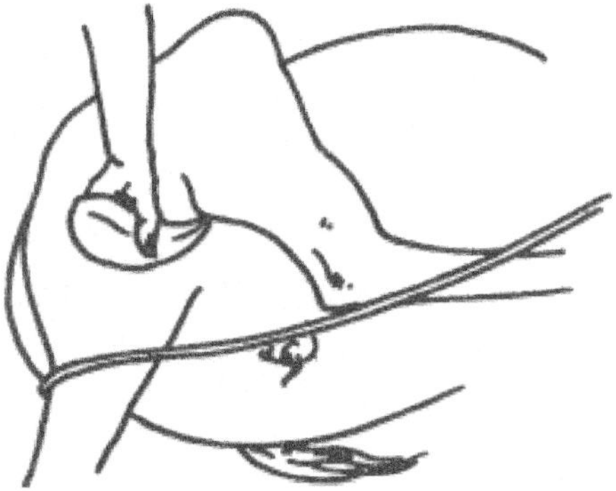

Castration method:

- Restrain the animal.
- Pull the testicles down gently.
- Hold the spermatic cord to one side by rolling it between the thumb and the forefinger.

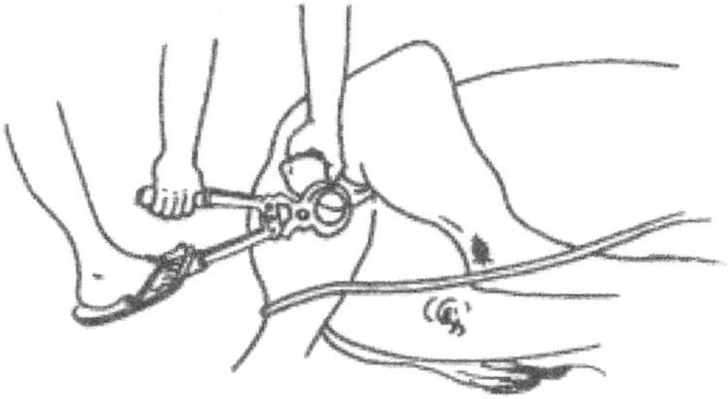

- Place the instrument with the teeth on the upper side.
- Place the spermatic cord inside the instrument.
- Close the jaws of the instrument all the way until a "click" sound is heard.
- Leave the instrument closed for 10 to 15 seconds.

- Open the instrument and remove it from the scrotum.
- Replace the instrument a little lower on the scrotum and crush the cord a second time.

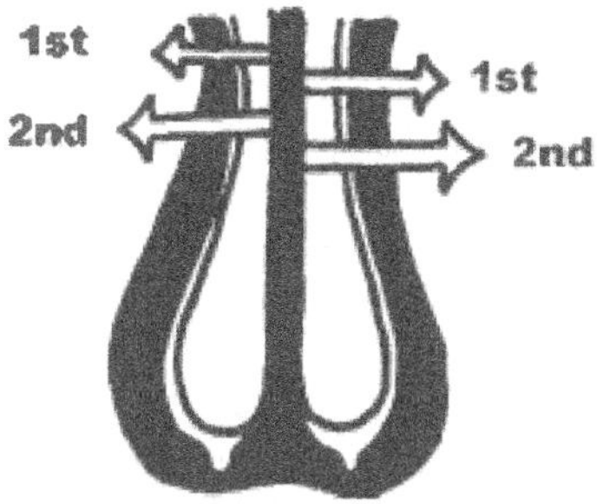

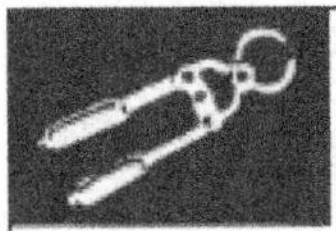

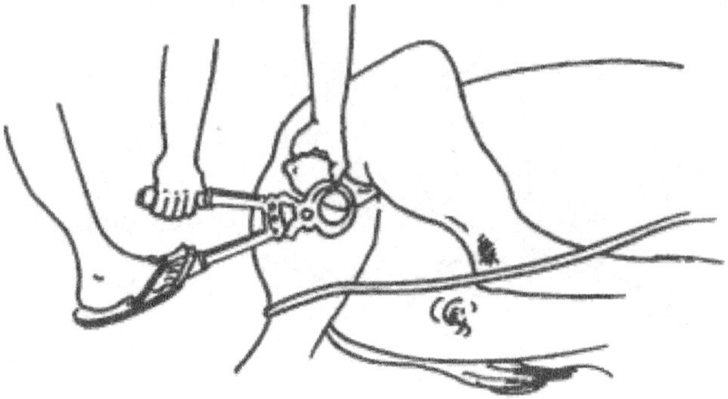

- Repeat the procedure on the opposite side of the scrotum, crushing the cord twice.
- Apply Himax® ointment to the crushed sites.

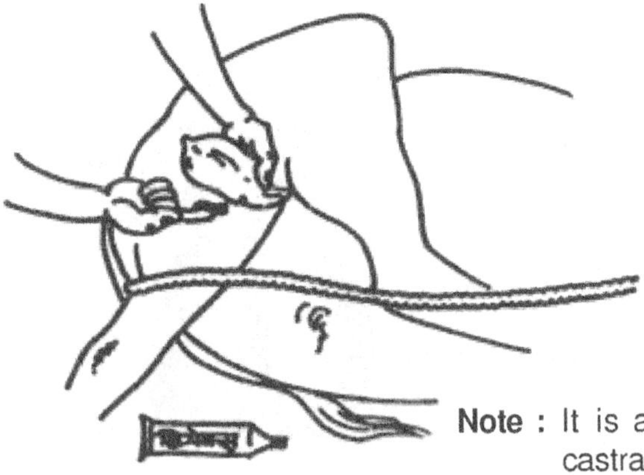

Things to remember:

- Let the animal rest for 5 days after castration.
- Keep the animal in a dry, clean place.
- The scrotum should be kept clean. There should not be any water or dirt on it.
- Give the animal nutritious feed.

Note : It is advisable to give penicillin injections the day before castration, the day of castration, and the day after to prevent infection - especially tetanus.

Trimming the Horn

Trimming the horn is done when the horn is overgrown, or to cure an infection in a horn.

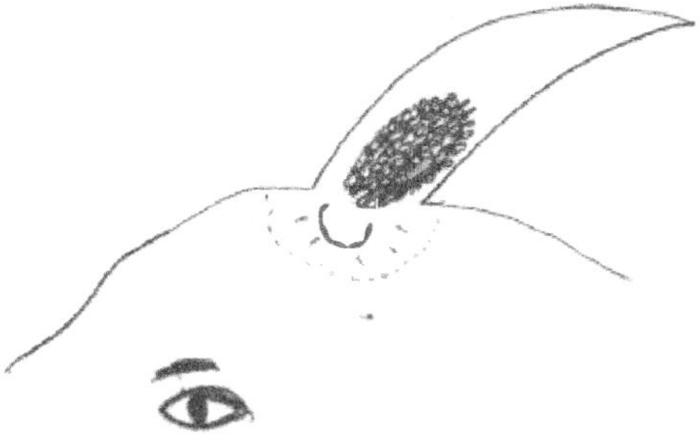

Why?

- To avoid body injuries.
- To treat infection in a horn.
- To make an animal docile.
- To give an animal a better look.

When?

We can remove horn as and when required.

How?

- There are several ways of trimming e.g. using a saw or a dehorning wire.
- While trimming, care should be taken to not damage blood vessels and tissues inside the living part of the horn.
- Measure the total length of the horn and cut only the upper one-fourth part.
- When the horn is too short, cut only a small portion of the horn.
- In case of infection in a horn, we need to remove the whole horn. At this time, you should get the assistance of a veterinarian.
- Always have the animal well restrained.
- In case of bleeding, powder may be used.

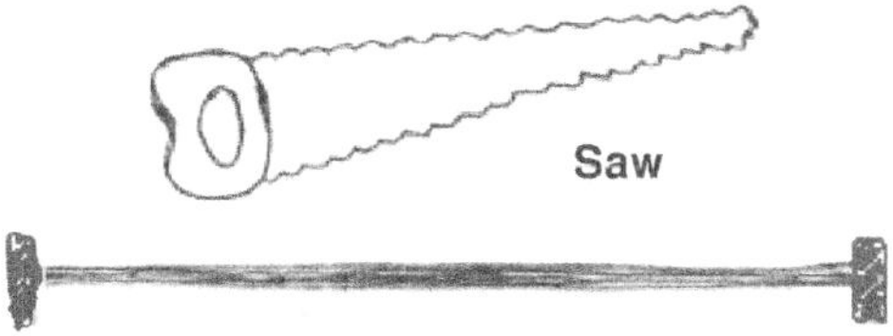

Hoof Trimming in Goats and Sheep

The process of cutting the overgrown hoof of an animal is known as hoof trimming. Overgrown hooves are mostly seen in stall-fed animals.

Why?

- To make movement easier.
- To reduce the risk of forming a wound or abscess inside the hoof.
- To prevent Foot Rot.
- To improve the looks.

When?

We can trim a hoof as and when required in large animals, but in small animals it is better to trim with an interval of two months.

How?

A sharp knife or a special hoof cutter can be used to trim a hoof. Special care should be taken to not cut the muscle and blood vessels inside. All sides of the hoof should be equally trimmed.

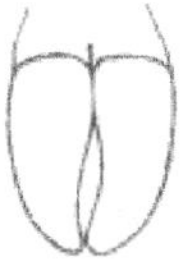

Disbudding

Disbudding is the arrest of horn growth at an early age, when the horn root is in the form of a bud or button. This is done when calves are within 1-2 weeks of age.

Why?

- To make an animal docile.
- To avoid injury during fighting.

How?

There are two methods of disbudding:

1. Hot iron rod
2. Caustic Soda or Caustic Potash

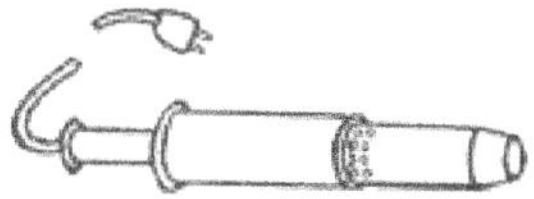

Hot iron rod:

- Specially designed iron rod is heated on a fire, or by electricity in the case of an electric dehorner.
- Apply the hot iron rod over the horn bud for a few seconds. This way the bud gets damaged permanently and will not grow.
- To prevent infection around the horn apply antibiotics and antiseptic cream.
- This method is bloodless and may be used in any season of the year.

Caustic Soda or Caustic Potash:

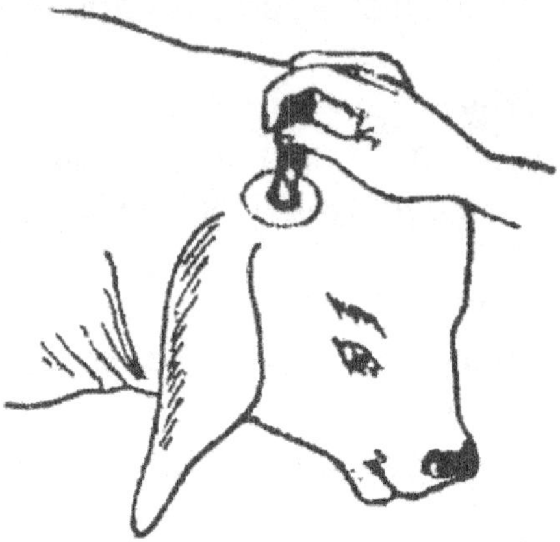

- Clip hairs around the horn bud and apply greasy Vaseline to protect the eye from being contaminated with the chemicals.
- Caustic Soda and Potash is available in the form of a stick, paste, or lacquer.
- Wrap the stick with aluminium foil and hold it in your hand.
- Dip the end of the stick in water and rub it over the bud in a circular motion.
- Stop rubbing when bleeding starts.
- Then apply vinegar over the bud.

Caution:

- Always do disbudding in the evening after the calves have finished suckling milk. This prevents the chemicals effecting the udder of the dam.
- Store the chemicals in a bottle with a lid.
- Don't use too much chemicals.

Manure Management

The proper utilization of animal waste (urine and faeces) in a proper way is known as manure management.

Why?

1. It makes the environment clean and healthier.
2. The materials which otherwise go to waste can be utilized.
3. It makes soil more fertile. It will add nutrients to soil, and increase the water-holding capacity of soil. It will help different useful bacteria and earthworms to grow in the soil.
4. It is a cheap and easy source of fertilizer.
5. It costs less than using inorganic fertilizer.

How?

A. Urine Management

B. Manure Management

A. Urine Management

1. A pit has to be constructed near the shed in such a way that only urine gets collected in it.
2. Dig a canal from the shed to the pit. It should be remembered that the pit should always be at a lower height than the shed.
3. The pit has to be covered with wooden block, metal sheet, or tree branches.
4. Collected urine can be used in the field by mixing with an equal amount of water OR can be transferred to the pit where solid faeces are being collected.

B. Manure Management

1. Construct a large pit near the shed with a roof or, when the pit is filled, cover the manure with plastic, straw, or leaves.
2. Then collect the faeces daily from the shed and store it in the pit.
3. Along with faeces, leaves, straw or litter can be stored.
4. At first, put the manure only in one side of the pit, as seen in fig. 1, and when it is filled cover it with mud, leaves, or plastic.
5. Again start collecting manure, put this in the other half of the pit.
6. After 2-3 months, use the manure as fertilizer.

Fig. 1

How to dig the pit?

- The size of the pit depends on the number of livestock and how much manure is needed.
- For urine collection, a pit of 1 m. in length x 0.5 m breadth and 1 m. depth is required for rectangular pit.
- For a circular pit, the diameter should be 1 m. and 1 m. in depth.

Fig. 2

Artificial Insemination

The process of depositing artificially collected semen from male animals in female genitalia by different artificial equipment is known as artificial insemination.

Why?

- Semen from exotic breeds can be used to inseminate the local breed to increase the productivity of offspring.
- Semen is diluted for artificial insemination (AI), so that a greater number of females can be bred compared to natural service.
- There is no need to keep a bull, so costs will be reduced.
- Semen can be stored, when frozen, for many years after the bull is dead.
- Sexually transmissible diseases can be controlled.
- Heavy and injured bulls can be used, still utilizing their breed advantages.
- It overcomes the difficulty of size and weight.
- The service will be available at home.

Disadvantage

- There are few disadvantages like; well trained practitioners are required and liquid nitrogen is needed to store the semen.
- Single insemination may not help the animal to become pregnant in many cases.
- Big bull's semen put into small females leads to bad birthing problems.

Required Equipment

- Semen Straw kept in Liquid Nitrogen.
- AI Gun
- AI Sith
- AI Glove
- Scissor
- Thermometer
- Lubrication

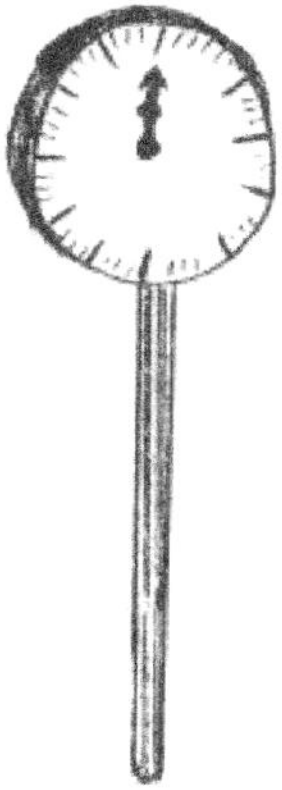

Technique

1. Preparation of the Straw

- Pick out semen straw from Liquid Nitrogen tank and keep it in a glass containing luke warm water for 30-60 seconds. The temperature of water should be 37°C (99°F). This helps to activate the semen within the straw.
- Then put the straw inside the AI gun and cut the top of the straw. Then cover it with the AI sith.

2. Preparation of the Animal

- Restrain Animal
- Clean the rear part
- Properly detect the heat and decide the time to inseminate.

3. Process

- Insert left hand into rectum to locate and grab the cervix.
- Then insert the AI gun into female genitalia and then insert its tip into the cervix.
- Then deposit the semen in the uterine horn or uterus.

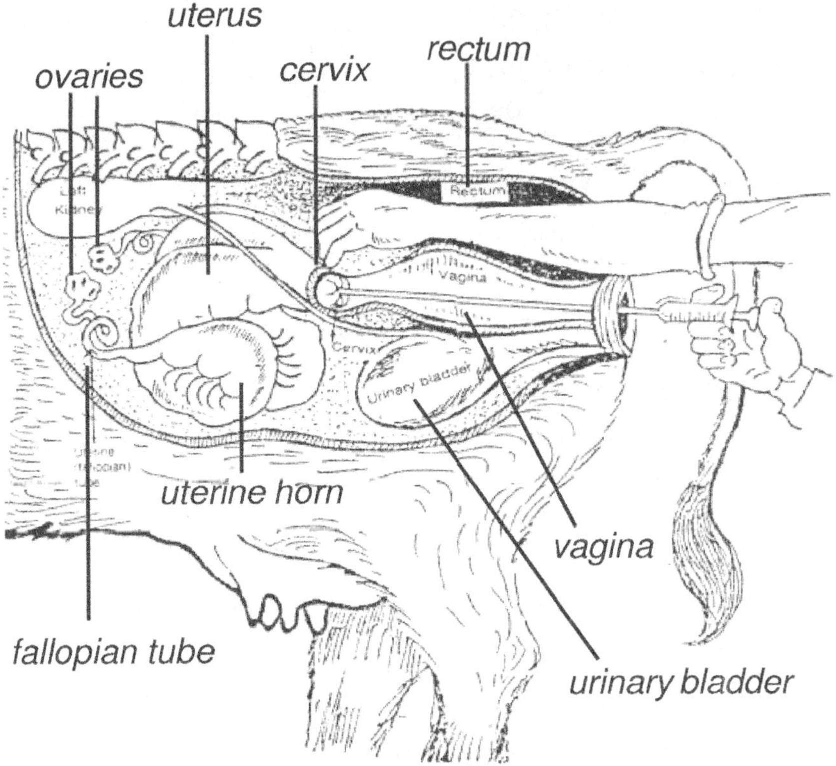

For AI to be successful, the female should be inseminated at the right time. The heat detection should be done carefully. When female comes in heat, AI should neither be done too early nor too late. Generally, it is done 12-16 hours after the heat detection.

Reporting to the Local Authorities

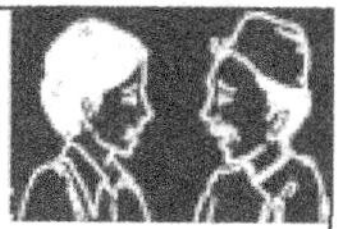

Why do we do this?

- To help prevent the spread of contagious disease.
- To raise awareness about the outcome of contagious diseases that are occurring in our village.
- To make use of the help available at the Government Livestock Centre or Sub-centre.

How do we report?

- First, examine the sick animals.
- Write down the signs of the sick animals in the record book.
- Determine if the disease is a "reportable disease".
- Discuss your findings with the community and give appropriate advice.
- Inform the local or livestock health authorities about the disease. Take your record copy to the Centre so you can discuss your findings with the doctor on duty.

Summary:

1. Examine sick animal

↓

2. Write findings in medical record notebook

↓

3. Determine if the disease is reportable. If so...

↓

4. Discuss with community members

↓

5. Report to Animal Health Care Centre or other local authority

Awareness Raising

What is awareness raising?

Teaching your friends and neighbours about a subject that is new to them.

Why do we do it?

- To motivate our community to improve their animal husbandry practices.
- To protect our animals from suffering and untimely death.
- To increase animal production in our community.
- To properly utilize local resources and materials.
- To transfer skills and knowledge to others.
- To improve the quality of life for our friends and neighbours.

What materials can we use?

- Posters
- Flip-charts and Flash cards
- Handouts

How do we use them?

Posters:

A poster is a picture with a specific message. This must be posted in a safe place where everyone can see it. It should be easy to read and easy to understand. It should not have too much writing on it.

Flip charts and flash cards:

This is a method of providing information to a small group in the form of pictures and words.

- Make sure the cards are shown in the right order.
- Each page is important. Give information about each page.
- Ask questions to determine if you are being understood.
- At the end, summarize the key points.

Handouts:

- A handout is a written paper with information about a particular subject.
- The handout should be in simple language, according to the education level of the readers. Do not try to write too much on one handout. Keep it simple.
- Make sure each member of the group has their own handout.
- After distributing the handouts, read it aloud. Or, you can ask a group member to read it aloud to the entire group.

Appendix 1

Sample to be taken specific to the different diseases:

S.N	Disease	Antimortem Sample	Post Mortem Sample
1.	Haemorrhagic Septicaemia	• Smear prepared from discharge from affected part of neck	• Pieces of Heart, Lungs, Liver, Intestine, Spleen
2.	Anthrax	• Blood smear from tip of ear	• Pieces of Heart, Spleen
3.	Brucellosis	• Blood and serum sample • Milk sample • Vaginal discharge	• Pieces of Intestine and liver from aborted Fetus • Placenta • Fetus if it is small in size
4.	Black Quarter	• Smear prepared from discharge in the swollen part	• Piece of muscles in the swollen area
5.	Tetanus	• Smear prepared from the wound	• Spinal Cord, Brain and Muscles
6.	Enterotoxaemia	• Blood sample	• Small intestine along with its contents
7.	Mastitis	• Milk or Blood sample	• Blood and Milk sample • Pus oozing out from teat.
8.	Foot and Mouth disease	• Contents of vesicle in mouth	• Kidney, Heart • Lymphnode
9.	PPR / RP	• Blood and Serum sample	• Lymphnode • Spleen and Lungs
10.	Rabies	• Saliva, Tear or Occular swab	• Whole head • Lymphnode
11.	Diarrhea	• Feces	• Pieces of Intestine
12.	Swine Fever	• Blood obtained directly from heart	• Blood obtained directly from heart • Spleen and Lymphnode.

Appendix 2

General Information about Livestock

Species	Temp °F	Temp °C	Pulse (per min.)	Respiration rate (per min.)	Gestation period (days)	Length of Estrus cycle	Length of Estrus
Cattle	100.5-102.5	38.0-39.2	50-70	35-45	280 ± 5	21 days	24 hours
Buffalo	101-102	38.0-39.0	40-60	35-45	310 ± 5	21 days	24 hours
Goat/Sheep	101.5-103.5	38.5-39.7	70-80	15-30	150 ± 5	21 days	24-36 hours
Swine	101.5-103.5	38.5-39.7	80-100	40-50	114 ± 5	21 days	72-192 hours
Horse	100-101	37.3-38.3	30-40	10-14	333 ± 5	21 days	72-192 hours
Poultry	106-107	41-42	350-450	12-40	-	-	-
Dog	100-102.5	37.3-39.2	85-105	15-30	62 ± 3	6-7 months	4-21 days (9 days in average)
Rabbit	102.1-104.1	39.2-40.0	250-300	30-60	32 ± 2	1 month or more	after mating
Cat	100-102.5	37.3-39.2	120-140	15-30	63 ± 2	15-21 days	9 days

Conversion Information

LENGTH
Kilometer (km) = 1000 meters
Meters (m) = 100 centimeters
Centimeter (cm) = 10 millimeters
1 Km = 0.62 miles
1m = 1.09 yards = 3.28 feet
1 cm = 0.394 inch

MASS
Kilogram (kg) = 1,000 grams
gram (g) = 1,000 milligrams (mg)
metric ton (t) = 1,000 kg
quintal = 100 kg
1 kg = 2.205 pounds
1 pound = 0.454 kg

VOLUME / CAPACITY
kiloliter (kl) = 1,000 liters
liter (l) = 1,000milliliters
1 gallons = 3.785 l
1 teaspoon = 5 cubic centimeters or 5 milliliters
3 teaspoons = 1 tablespoon
1 tablespoon = 15 cc/ml

C	F
40.6	105.1
40.4	104.7
40.2	104.3
40.0	104.0
39.8	103.7
39.6	103.3
39.4	102.9
39.2	102.6
39.0	102.2
38.8	101.8
38.6	101.5
38.4	101.2
38.2	100.8
38.0	100.4
37.8	100.1

C	F
37.6	99.7
37.4	99.3
37.2	99.0
37.0	**98.6**
36.0	96.8
35.0	95.0
34.0	93.2
33.0	91.4
32.0	89.6
31.0	87.8
30.0	86.0
29.0	85.2
28.0	82.4
27.0	80.6

MEDICINE BOOK FOR VILLAGE ANIMAL HEALTH WORKERS

How to use this book

To get the most benefit from this book, please read the following suggestions on how to use it.

The table of contents is arranged into sections to help you find the information quickly. After looking up the name of the medicine, its page number is located to the right. There are seven sections, A through G, arranged as follows:

Medicines for treatment of fever
Medicines for treatment of internal parasites
Medicines for treatment of external parasites
Medicines used for treatment of digestive problems
Vaccines against contagious diseases
Medicines used for treating eye problems
Miscellaneous medications

Make sure you read all the information before trying to use a particular medicine. For each medicine, information is given about its name, other common names, reasons for its use, forms available, and methods of use. Clear numbers are given so you will know exactly how to use the medicine.

On the last page of the book, a simple technique for measuring the dose of powdered medicine is given. Some types of medicine may have many different trade names (trade names are marked by the ® sign in this book), yet inside the bottle or packet, the ingredients are the same. It is important to know that all these medicines work in the same way.

- If you cannot find the name of a medicine in this book, the medicine shop owner can help you. When buying a new medicine that is not in this book, make sure you get the following information from the medicine shop owner:

    a. Name of medicine
    b. When to use it (what disease it is for)
    c. Dose
    d. How many times per day (or per week) the medicine should be given
    e. Precautions
    f. How to mix and use
    g. What other forms of medicine are available
    h. Expiration date

- After examining a sick animal, you can use the information in this book and in the VAHW textbook to help you decide which medicine to use. We recommend that you always keep a written record of each animal you examine. Always make a written record of the kind and amount of medicine used.

- This book is designed to be used as a reference book for someone who has successfully completed Village Animal Health Worker training. It should not be used by an untrained person.

Animal Health Training and Consultancy Service

Ramghat, Pokhara, Nepal

Table of Contents

Table of Contents

Table of Contents

Table of Contents

G) Miscellaneous medications

A

Procaine Penicillin

When do we use it?

- Fever
- Pneumonia
- Dystocia
- Retained Placenta
- Deep infected wounds
- Mastitis

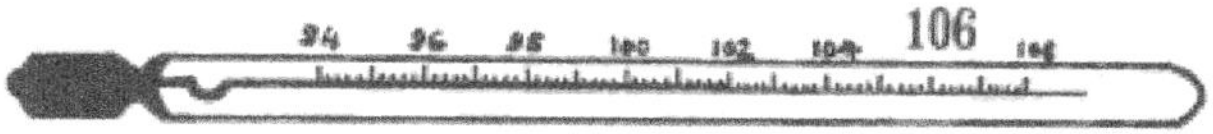

Dose : The dose of procaine penicillin is measured in *units.*

Cattle, Buffaloes:
4,000,000

Sheep, Goats:
1,000,000

Horses:
3,000,000

Pigs: 1,000,000

Dogs: 500,000

Cats: 200,000

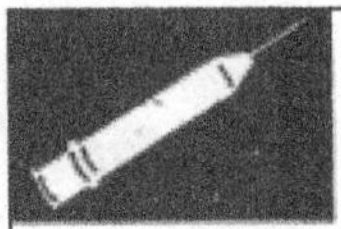

Precautions:

- Penicillin injections should not be given at the same time with Terramycin® injections.
- Penicillin injections should be given for a minimum of 3 days in a row.
- Once the powdered penicillin has been mixed with water, it can be used for 1 week. Store it in a cool, dark place.
- Do not consume the milk from a treated animal for a least 1 to 2 days after giving the last dose.

How many times do we give the dose?

- Give 1 dose each day.
- Penicillin should be given each day for at least 3 days in a row.
- It can be given for 5 days in a row if necessary.

How do we give it?

- Inject in the muscle (intramuscular, IM)

What forms of the medicine are available?

Powdered form in small glass vials. Vials are available containing 4 million, 2 million, or 400,000 units of procaine penicillin. Add 7ml of sterile water to make a solution.

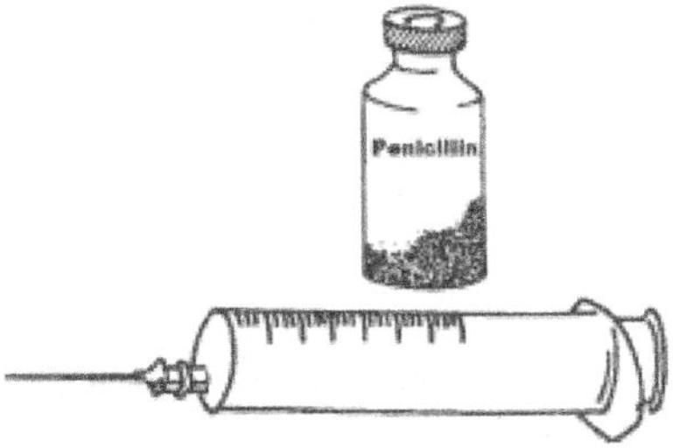

A

Benzathine Penicillin, Penidur®

When do we use it?

- Fever
- Pneumonia
- Dystocia
- Retained placenta
- Deep or infected wounds
- Mastitis

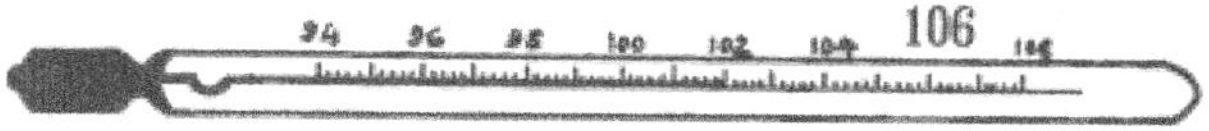

Dose: The dose of Penidur® (benzathine penicillin) is measured in *units.*

Cattle, Buffaloes, Horses:
2,400,000

Large Sheep, Goats, Pigs,
Dogs: 1,200,000
Small Sheep, Goats, Pigs,
Dogs: 600,000

Precautions:

- Procaine penicillin can be given with benzathine penicillin, but Terramycin® should not be given at the same time as benzathine penicillin.
- Do not consume the milk from a treated animal for a least 1 to 2 days after giving the last doze.

How many times do we give the dose?

- One dose of Penidur® lasts about 3 days in the animal. This makes it very useful if we cannot give procaine penicillin every day. Give 1 dose of Penidur® every 3 days.

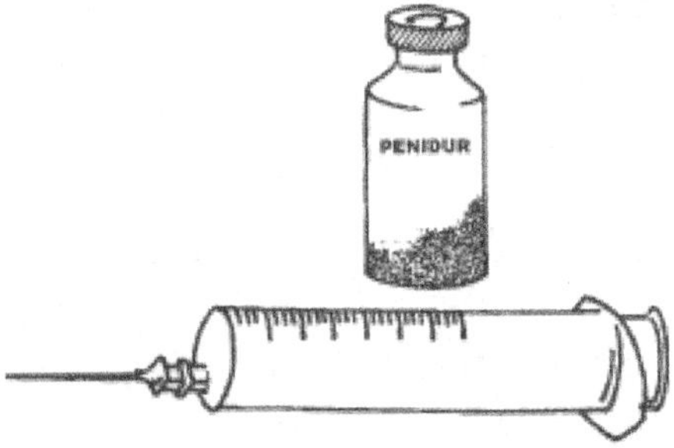

How do we give it?

- Inject in the muscle (intramuscular, IM)

What forms of the medicine are available?

- Powdered form in small glass vials. Vials are available containing 2.4 million, 1.2 million, or 600,000 units of benzathine penicillin. Add sterile water as per instruction given on the label.

A Oxytetracycline, Terramycin®, Telon®, Oxyvet®, Olicyclin®, Stecline®

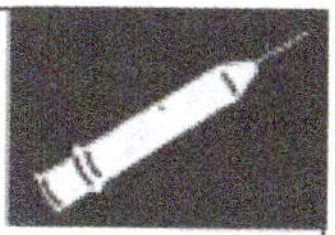

When do we use it?

- HS
- Pneumonia
- Dystocia
- Retained Placenta
- Deep or infected wounds
- Mastitis
- Fever

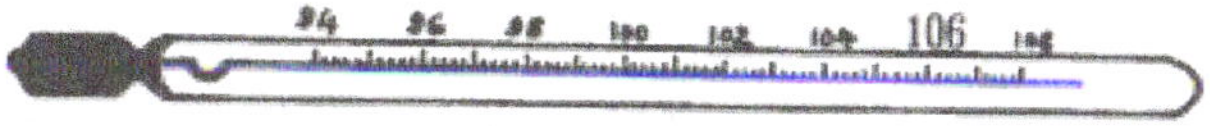

Dose: The dose of Terramycin® is measured in cc's or mls

Cattle and Buffaloes:
15 to 20 ml

Goats, Sheep, Calves:
3 to 5 ml

Pigs:
5 to 10 ml

Dogs and Cats:
1 to 2 ml

Precautions:

- Do not inject more than 20 cc at the same site in cattle and buffalo.
- Terramycin® injections can be dangerous in the horse and mule. It should not be used without a veterinary doctor's permission.
- Do not consume the milk from a treated animal for a least 3 days after giving the last doze.

How many times do we give it?

- One dose lasts about 24 hours.
- Give 1 dose per day and continue for at least 3 days in a row.
- If necessary, the medicine can be continued for up to 5 days in a row.

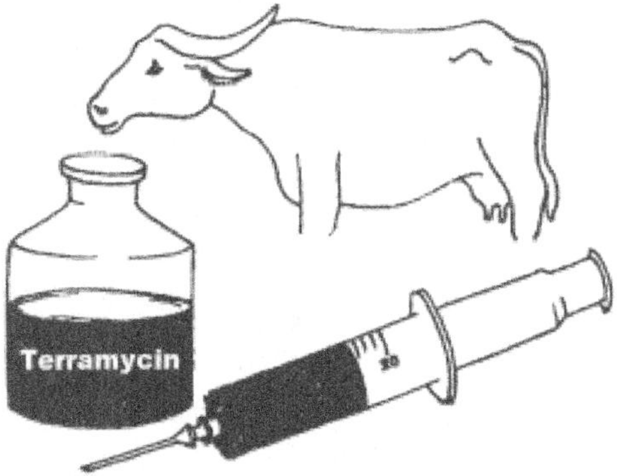

How do we give it?

- Inject in the muscle (intramuscular, IM)

What forms of the medicine are available?

Available in liquid form in 30 ml and 50 ml glass vials

Note: If the fever does not come down within 24 hours after giving Terramycin, then procaine penicillin should be tried to treat the fever.

A Oxytetracycline, Terramycin® L.A., OTC® L.A.

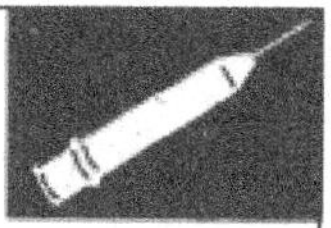

When do we use it?

- HS
- Fever
- Pneumonia
- Dystocia
- Retained Placenta
- Deep or infected wounds
- Mastitis

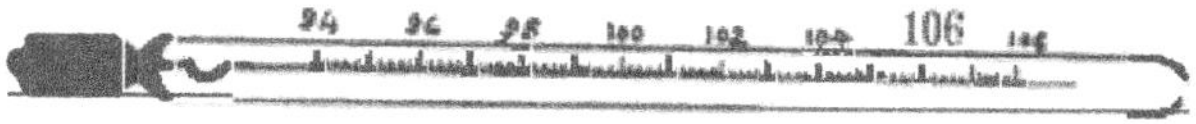

Dose: The dose is measured in cc's or mls.

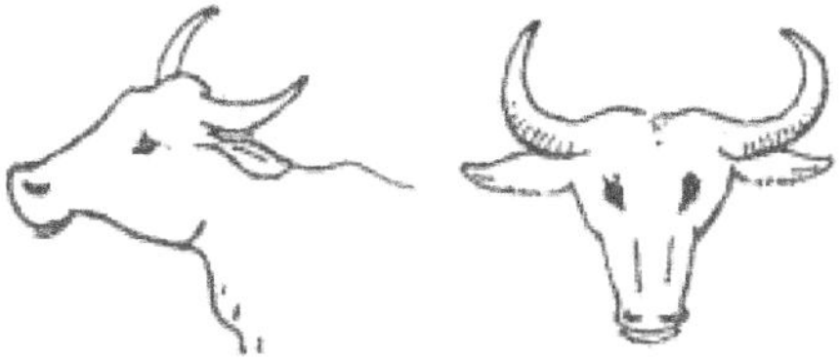

Cattle, Buffaloes:
40 to 50 ml

Sheep, Goats, Pigs:
5 to 10 cc's

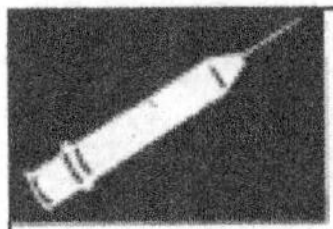

Precautions

- Do not consume the milk from a treated animal for at least 3 days after giving the last dose.
- Do not use in horses or mules.
- Do not inject more that 20 cc in one site in cattle and buffalo.

How many times do we give it?

- 1 injection will last for 3 days.

How do we give it?

- Inject in the muscle (intramuscular, IM)

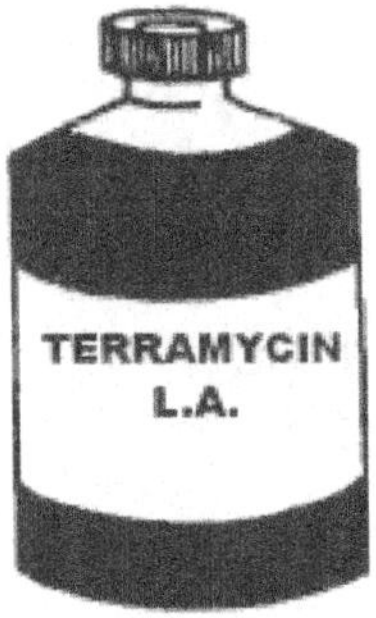

What forms of the medicine are available?

- Liquid in 10 ml and 30 ml glass vials.

Sulfadimidine, Sulfamethazine, Diadin®

When do we use it?

- Coccidiosis
- Fever in pigs
- Dystocia
- Diarrhoea in poultry, kids, lambs, pigs
- Retained placenta
- Uterus infection
- Deep infected wounds

Dose:

For use in the uterus:

Cattle, Buffaloes:
2 boluses

Goats, Sheep:
½ bolus

For oral use:

Calves, kids, pigs, dogs:
1 bolus per 50 kg body weight

Poultry:
2 boluses per 100 birds per day mixed in the drinking water

Precautions:

- Make sure the animals have adequate drinking water.
- This medicine does not work well if given orally to adult cattle, buffalo, sheep, and goats.

How many times do we give it?

- On the first day, give it twice daily for calf, kid, pig, and dog.
- On the 2nd through 5th day, give once per day.

How do we give it?

- Feed the bolus hidden in leaves, flour, or mixed with the drinking water.
- For poultry, dissolve boluses in drinking water.
- For uterus problems, boluses should be placed as far into the uterus as you can reach.

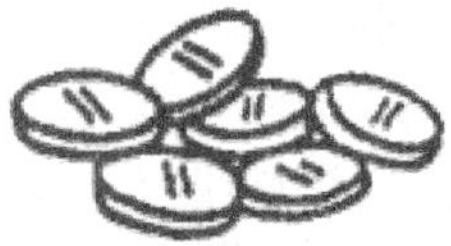

What forms of the medicine are available?

- 5 gram bolus

A

Herbina®

When do we use it?

- Fever
- Respiratory disorders

Dose:

Cattle, buffaloes, horses:
2 capsules

Goats and Sheep:
1 capsule

Calves:
1 capsule

Precautions:

- The medicine should be kept in an airtight bottle.

How many times do we give it?

- Twice daily for 3 days.

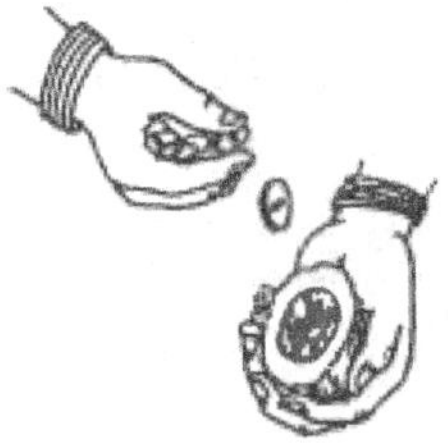

How do we give it?

- Feed with leaves, fruit, or mixed with flour.

What forms of the medicine are available?

- Bottles of 12 capsules and bottles of 6 capsules.

A

Herbal Anti-pyretic & Anti-inflammatory Feverdon®

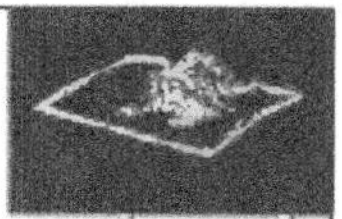

When do we use it?

- Fever
- Urine retention
- Oedema

Dose:

Cattle, buffaloes, horses:
60 to 100 grams

Sheep and goats:
30 to 60 grams

Calves:
30 to 60 grams

Precautions:

- Keep medicine in an air tight package.

How many times do we give it?

- Two times daily.
- Can be given up to 3 days in a row.

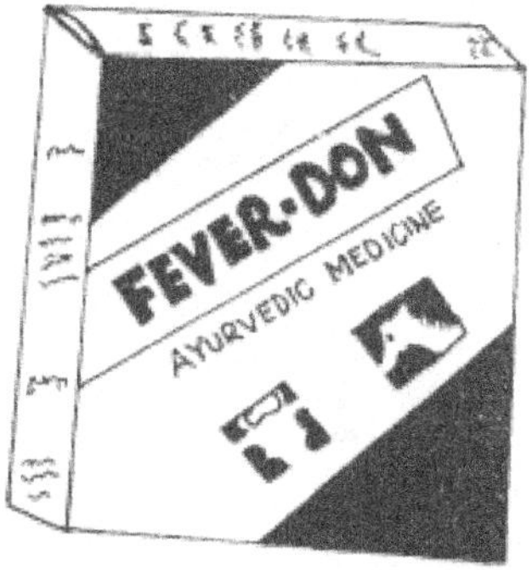

How do we give it?

- Mix with a bucket of water and give with a drenching bottle or tube.

What forms of the medicine are available?

- 200 gram and 1 kg packets.

B

Oxyclozanide, Distodin®, Hexanide®, Clozan®, Okazan®

When do we use it?

- For the treatment and prevention of liver fluke.

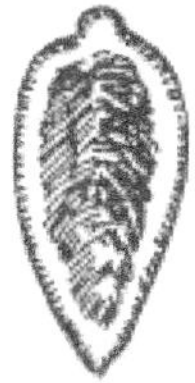

Dose:

Give one bolus for 100 kg of body weight, for example:

Improved buffaloes:	5 boluses (1g)
Local buffaloes:	4 boluses
Improved cattle:	4 boluses
Local cattle:	3 boluses

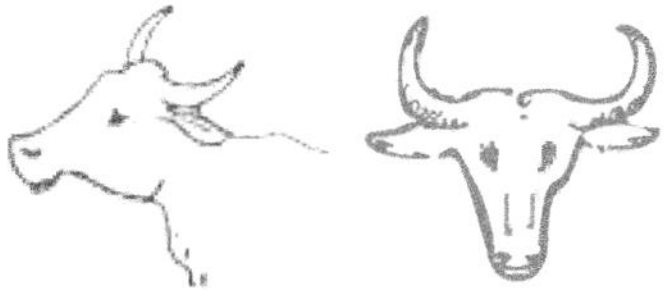

One tablet per 20 kg of body weight, for example:

Improved goats, sheep:	2 tablets (200mg)
Local goats, sheep:	1 tablet

Precautions:

- Do not give during the first and last month of pregnancy.
- Do not consume the milk from a treated animal within 24 hours of giving this medicine.
- Do not consume the meat from a treated animal for 14 days after giving this medicine.

How many times do we give it?

- Give the entire dose at one time to a healthy animal.
- If the animal is weak, give Vitamin B injections once daily for 3 days, then split the dose into 2 parts. Give the first part on the 4th day and the second part on the 6th day.
- Give the medicine every 6 months from 3 months of age.

How do we give it?

- Mix the medicine with tasty leaves or flour.
- Crush the bolus or tablet, mix with water and give with drenching bottle or tube.

What forms of the medicine are available?

- 1 gram bolus: 4 per foil strip
- 200 mg tablet: 10 per foil strip

Note: 1 bolus equals 5 tablets.

B

Fenbendazole, Panacur®, Panfugal®

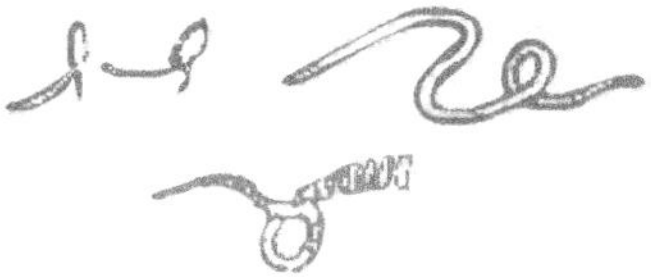

When do we use it?

- For the treatment of small roundworms.
- For the treatment of large roundworms.
- For the treatment of tapeworm in sheep and goats.

Dose:

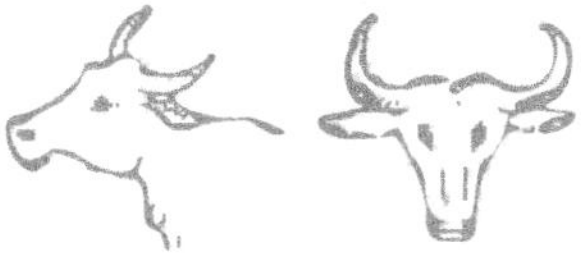

Animal	Tablet (150 mg)	Liquid
Large cattle:	15 tablets	150 ml
Medium cattle:	10 tablets	100 ml
Small cattle:	5 tablets	50 ml
Large Sheep, Goat	2 tablets	20 ml
Small Sheep, Goat	1 tablet	10 ml

- In the case of horses, kids, lambs, and pigs: give ½ (150mg) tablet or 5 ml for each 15 kg of body weight.

- For dogs, give 6 (150mg) tablets or 60 ml per 10 kg of body weight.

Precautions:

- Other sizes of tablets may be available. Make sure you have the correct size of tablets!

How many times do we give it?

- For cattle and calves, give every 6 months.
- For young sheep, goats, and dogs (under the age of 1 year), give every 3 months. Give medicine to adult animals every 6 months.
- For pigs, give every 6 months.

How do we give it?

- Tablet can be fed with grass, tasty leaves, or mixed with water and drenched.
- For pigs, mix in with their feed.
- Mix 6 grams of powder with 100 ml of clean water to make the medicine solution.

What forms of the medicine are available?

- Small tablets of 150 mg
- Large tablets of 1.5 grams may also be available. One large tablet equals 10 small tablets.
- Powder form is also available.

B

Piperazine, Piperacite®

When do we use it?
For treatment of large round worms.

Dose:

Calves (below 6 months of age):
For 1 kg body weight - 1 tablet (300mg)

Pigs:
For 1 kg body weight - 1 tablet (300mg)

Horses:
For 4 kg body weight - 1 tablet (300mg)

Adult Chickens:
1 tablet (300mg)

Dogs :
For 2 kg body weight - 1 tablet (300mg)

Precautions:

- Not necessary to use in calf once it has reached 6 months of age.
- Not necessary to use in sheep and goats.

How many times do we give it?

- Calf and dog: Give at 3, 7, and 16 weeks of age.
- Adult dog: Give every 6 months.
- Pigs, poultry & horse : Give every 3 months.

How do we give it?

- Tablet: mix with tasty food OR crush and mix with water and give with a drenching bottle or tube.
- Liquid: drench with bottle or tube.
- For poultry, mix in drinking water.

What forms of the medicine are available?

- Many forms of tablets and liquids are available. See the chart on the next page.

Forms and Doses of Piperazine Medication

Animal	Piperazine 300 mg Tablet	Piperazine 62.5% Liquid	Piperazine Hexahydrate 46% Liquid	Piperazine Citrate 15% Liquid
Calves: (0 to 6 months of age)	1 tablet per 1 kg body weight	2 ml per 5 kg body weight	5 ml per 10 kg body weight	8 ml per 5 kg body weight
Pigs	1 tablet per 1 kg body weight	2 ml per 5 kg body weight	5 ml per 10 kg body weight	8 ml per 5 kg body weight
Dogs, Cats, Chickens:	1 tablet per 2 kg body weight	1 ml per 2 kg body weight	1 ml per 4 kg body weight	3 ml per 5 kg body weight

Oxyclozanide, Tolzan®, Okazan®

When do we use it?
For treatment of liver flukes.
For prevention of liver flukes.

Dose:

Improved buffaloes:	100 ml
Local large buffaloes:	90 ml
Local small buffaloes:	75 ml

Local large cattle:	75 ml
Local small cattle:	60 ml

Local large sheep and goats:	10 ml
Local small sheep and goats:	7 ml

CAUTION!

Precautions:

- Do not consume the milk from a treated animal within 24 hours of giving this medicine.
- Do not consume the meat from a treated animal for 14 days after giving this medicine.
- Do not give this medicine during the first and last month of pregnancy.

How many times do we give it?

- Give entire dose at one time.
- Give the medicine every 6 months from 3 months of age.

How do we give it?

Mix with water and give with a drenching bottle or tube.

What forms of the medicine are available?

White liquid in bottles of 90 ml and 1 liter.

Albendazole, Valbazen®, Calbend®

B

When do we use it?

- To treat and prevent liver fluke
- To treat and prevent small roundworms, including lungworms

Dose:

Large cattle, buffaloes:	6 large tablets (600 mg)
Medium-sized cattle, buffaloes:	5 large tablets
Small cattle, buffaloes:	4 large tablets

Large sheep and goats: (150mg)	3 small tablets
Medium-sized sheep and goats:	2 small tablets
Small sheep and goats:	1 small tablet

Precautions:

- Do not eat the meat of a treated animal for 14 days after the last treatment.
- Do not drink the milk of a treated animal for 3 days after the last treatment.
- Do not give this medicine during the first and last month of pregnancy.
- There are 2 sizes of tablets. Make sure you use the correct size.

How many times do we give it?

- Give entire dose at one time.
- Give every 6 months from 3 months of age.

How do we give it?

- Can mix with grass or tasty leaves and feed to animal.
- Can mix with water and give with a drenching bottle or tube.

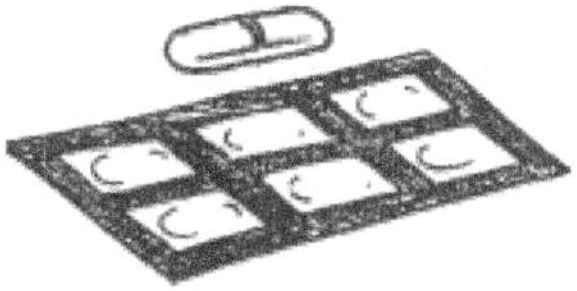

What forms of the medicine are available?

- Large tablets (600 mg) in foil strip of 6.
- Small tablets (150 mg) in a foil strip of 10.

Note: 1 large tablet is equal to 4 small tablets.

Delthamethrin, Butox®, Neocidol®

C

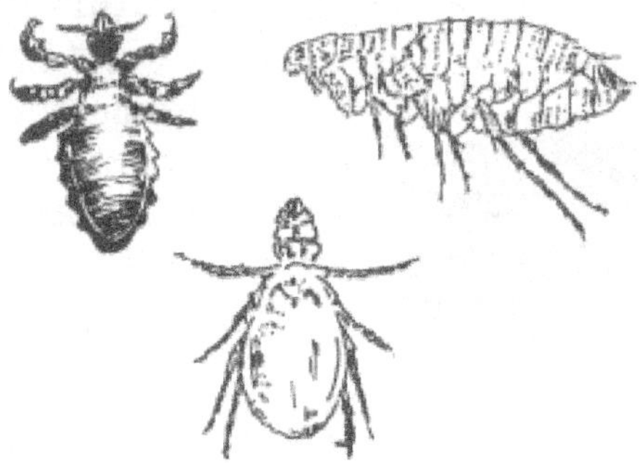

When do we use it?

- To treat an infestation of lice, fleas, or ticks.
- For the treatment of mange.

Dose:

As much as needed.

CAUTION!

Precautions:

This is a poisonous medicine! Keep it away from children. Cover animal's nose, eyes, and mouth with a cloth while using this medicine. Cover your eyes, nose and mouth also. Wash yourself thoroughly after handling this medicine.

How many times do we use it?

- Use once to treat lice, fleas, or ticks and then as needed.
- For treatment of mange, use once a week for two weeks in a row.

How do we use it?

- Add 1 part medicine, to 250 parts of water.
- For lice, flea, or tick infestation, apply the solution all over the animal's body.
- For mange, clean the affected area with soap and water and apply the solution.

What forms of the medicine are available?

- Liquid in 50 ml bottle
- Liquid in 100 ml bottle

Sulphur

C

When do we use it?
For the treatment of mange

Dose:
As much as needed.

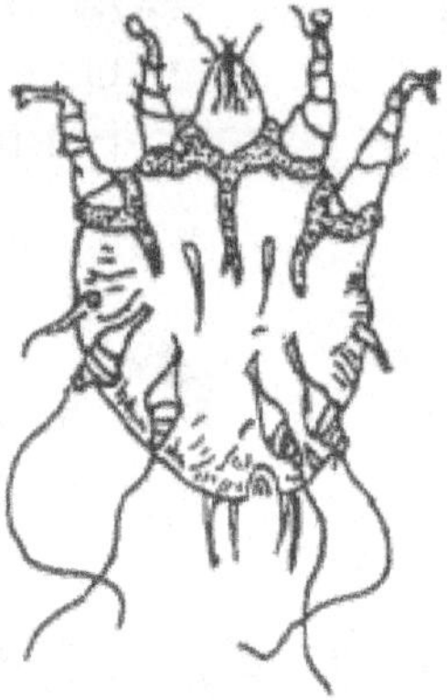

CAUTION!

Precautions:
Use only on the outside of the body. Do not give it by mouth.

How many times do we use it?
Once a week for 4 weeks in a row.

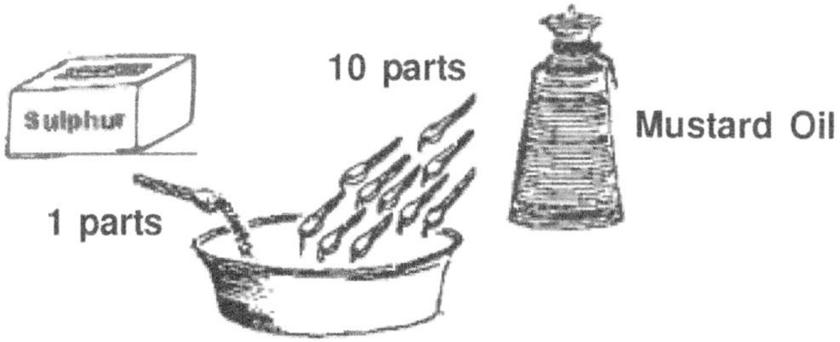

How do we use it?

1. Make a mixture of 1 part sulphur powder and 10 parts mustard oil.
2. Remove the dead skin and scabs with a rough cloth.
3. Wash the animal and dry it.
4. Apply the sulphur and oil mixture to the affected areas.

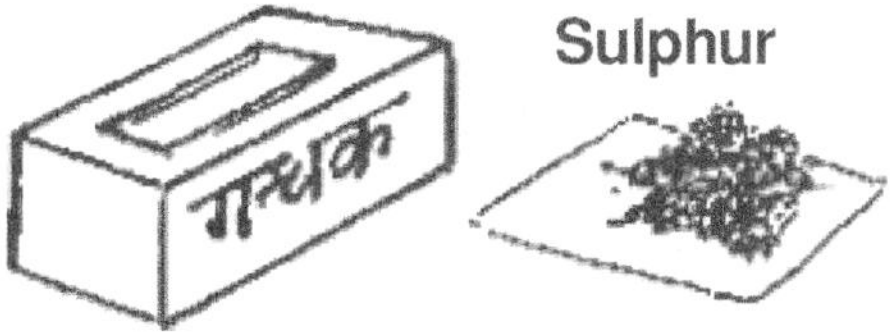

What forms of the medicine are available?

- Yellow powder in 450 gram (1 pound) packets.
- Block form is also available.

Tobacco Juice

When do we use it?
To treat an infestation of fleas, ticks, or lice.

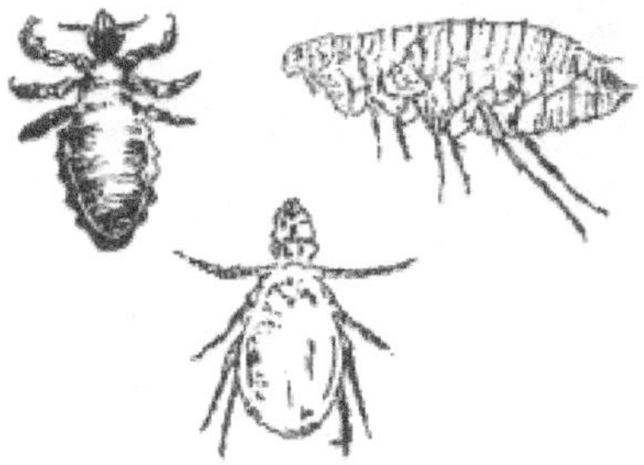

Dose:
As much as needed.

How do we prepare the tobacco juice?
- Chop the tobacco leaves into small pieces

Method 1:
- Mix the chopped leaves with 5 liters of water. Add a few small pieces of laundry soap.
- Boil the mixture until the volume is decreased to 2.5 litres.
- Filter the liquid and store in a bottle.

Method 2:
- Mix the chopped leaves with 5 litres of water. Add a few small pieces of laundry soap.
- Allow to soak for 24 hours.
- Filter the liquid and store in a bottle.

CAUTION!

Precautions:

- Keep the solution away from children.
- Cover your eyes, nose and mouth while using it.
- After using, wash yourself thoroughly.
- Keep doors and windows open during boiling of the mixture.
- Do not bathe the animal for 2 days after applying the liquid.

How many times do we use it?

- Use once for the control of fleas or ticks.
- Use once a week for the control of lice.

How do we use it?

- Mix 1 part of the concentrated solution with 1 part water and apply to the animal with a clean cloth.
- Apply the unboiled solution directly to the animal with a clean cloth.
- Make sure the entire animal is covered with the solution, but do not get it into the animal's eyes, nose, or mouth.

What forms of the medicine are available?

- Leaves
- Juices

Herbal Medicine
Himalayan Battisa, Herminsa®, Shilpa Battisa

D

When do we use it?

- Inappetence
- Stomach problems
- Not ruminating
- When deworming a weak animal

Dose :

Cattle, horses, buffalo, and mules:	50 grams	
Sheep and goats:	15 grams	
Large pigs and calves:	25 grams	 

Precautions:
There are many causes of inappetence. If possible, medicine should be given to treat the exact cause.

How many times do we give it?
2 times per day for 3 days in a row.

How do we give it?
Mix with a bucket of water and give with a drenching bottle or tube.

What forms of the medicine are available?
Brown powder in 100 gram, 400 gram, and 1 kg packets.

D

Herbal Medicine
Timpol®, Afron®, Tympanimal®

When do we use it?

To treat bloat from any cause, such as:

- When animal has eaten too much grain
- When animal has eaten rotten grass or feed.
- When animal has eaten too many young shoots and leaves.
- When the animal has eaten too much legume grass.

Dose :

Cattle, Buffaloes: 50 grams

Sheep, Goats: 30 grams

Precautions:
Keep the medicine stored away from air and sunlight.

How many times do we give it?
Every 4 hours until the animal has recovered.

How do we give it?

- Mix with warm water and give with a drenching bottle or tube. Mixing with warm water is more effective than mixing with cold water.

What forms of the medicine are available?
Brown powder in 100 gram and 1 kg packets.

Herbal Medicine, Neblon®

D

When do we use it?

- For treatment of diarrhoea caused by rotten grass or feed.
- For treatment of diarrhoea caused by coccidiosis.
- For treatment of diarrhoea caused by parasites.

Dose :

Large cattle, buffaloes:	50 grams
Small cattle, buffaloes, horses:	30 grams
Calves:	30 grams

Dose:

Foals:	6 to 10 grams
Adult pigs:	6 to 10 grams
Dogs and piglets:	2 to 3 grams
Sheep and goats:	15 grams

CAUTION!

Precautions:

- Always give plenty of water when giving Neblon®.
- For young animals, it is better if you can drench with a salt-sugar (rehydration) solution.

How many times do we give it?

Give every 12 hours until the animal recovers.

How do we give it?

- Drench with plenty of clean water or rehydration solution.
- Use a drenching bottle or tube.

What forms of the medicine are available?

Brown powder in 100 gram and 1 kg packages.

D

Mustard Oil

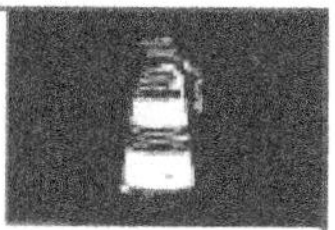

When do we use it?

Given by mouth to treat bloat from any cause, such as:

- When animal has eaten too much grain
- When animal has eaten rotten grass or feed
- When animal has eaten too much legume grass
- When animal has eaten too many young shoots and leaves

Mixed with sulphur

- Mixed with sulphur and applied to the skin for treatment of mange.

Dose:

Cattle, buffaloes: ½ to 1 liter

Sheep and goats: 100 ml

How many times do we give it?

- Give entire dose at one time.
- Dose can be repeated every hour until the animal is recovered.
- For mange treatment, refer to pages about use of sulphur.

How do we give it?

Give with a drenching bottle or tube.

What forms of the medicine are available?

- Liquid (range of sizes)

Turpentine Oil

When do we use it?

- To treat bloat in ruminant animals.
- To treat colic in horses.
- To kill maggots in a wound.

Dose:

Horses: 50 ml

Cattle and buffaloes: 50 ml

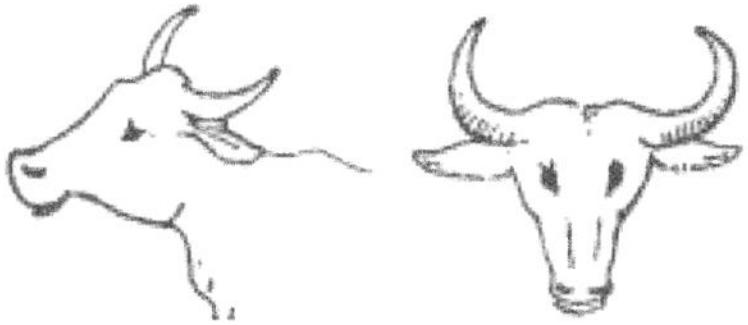

Dose:

Sheep and goats: 10 ml

CAUTION!

Precautions:
Always keep the bottle tightly closed.

How many times do we use it?

- For treatment of bloat and colic, it can be repeated every 2 hours until the animal improves.
- In a maggot-infested wound, it can be used as often as needed until all the maggots are gone.

How do we use it?

1. For cases of bloat and colic:
Add 1 part turpentine oil to 10 parts mustard oil, and give with a drenching bottle or tube.

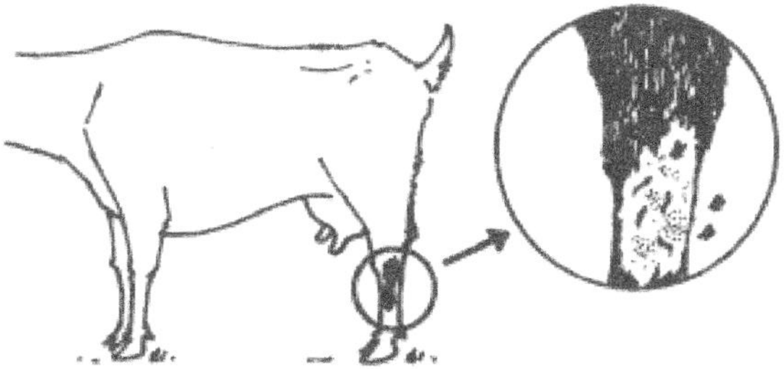

2. For treatment of maggot-infested wounds:
Soak cotton with turpentine oil and pack into wound for 5-10 minutes.

What forms of the medicine are available?
Clear, light yellow liquid in 120 ml glass bottles.

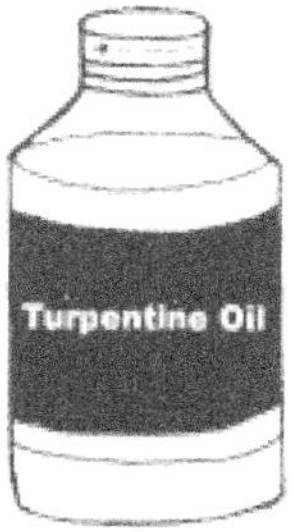

Yeast

When do we use it?

- Indigestion
- Coccidiosis in calves
- Change in feed or fodder
- Inappetence

Dose:

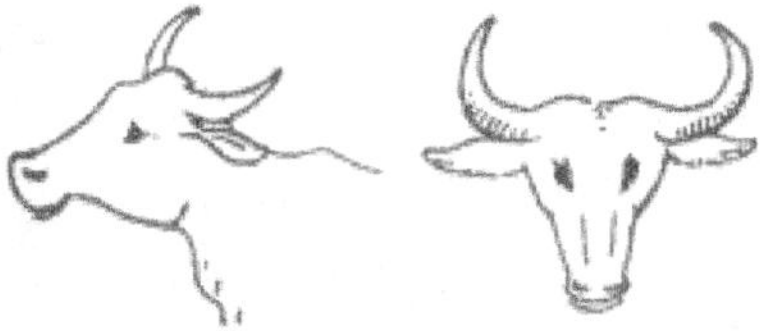

Cattle and buffaloes:	4 to 5 handfuls (of pellets/tablets)
Calves:	2 handfuls
Sheep and goats:	1 handful

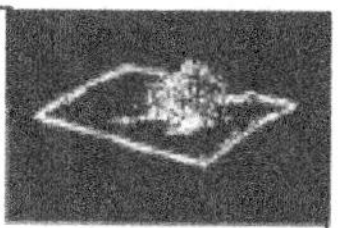

CAUTION!

Precautions:
Store in a dry place.

How many times do we give it?
Give once daily for 3 days.

How do we give it?

- Mix with grass or tasty leaves for adult animal
- For calf or kid, mix with water and give with a drenching bottle or tube.

What forms are available?
Locally made pellets and tablets.

Magnesium Sulphate

D

When do we use it?

- Intestinal blockage (impaction)
- Constipation
- Plant poisoning
- Bloat
- May be used in combination with other poisoning or bloat medicines.

Dose:

Cattle and buffaloes:	200 to 300 grams
Calves:	50 grams
Sheep and goats:	25 grams
Horses and pigs:	30 grams

Precautions:

- Always give plenty of water with this medicine.
- It is safe to use in a pregnant animal.

How many times do we give it?

- Dose can be repeated after 4 hours if needed.
- Do not give more than twice per day.

How do we give it?

Mix with warm water and give with drenching bottle or tube.

What forms of the medicine are available?

Coarse, white crystals in 400 gram and 450 gram packets.

Salt-Sugar (Rehydration) Solution

D

When do we use it?
- For the treatment of diarrhoea.
- For the treatment of inappetence.
- As temporary milk replacement for a nursing calf or kid.

Dose:
- Give as much as needed.
- An animal needs about 4 to 8 litres of water per 100 kg of body weight every day. It needs more if it is sick with diarrhoea.

How do we make it?
- 3 cups of water
- 1 pinch of salt
- ½ handful of sugar

How do we give it?
Give with a drenching bottle or tube

E

HS Vaccine

Also called HS Broth Vaccine

When do we use it?
To prevent Haemorrhagic Septicaemia (HS) disease.

Dose:

Cattle and buffaloes: 55 cc's (5 mls)

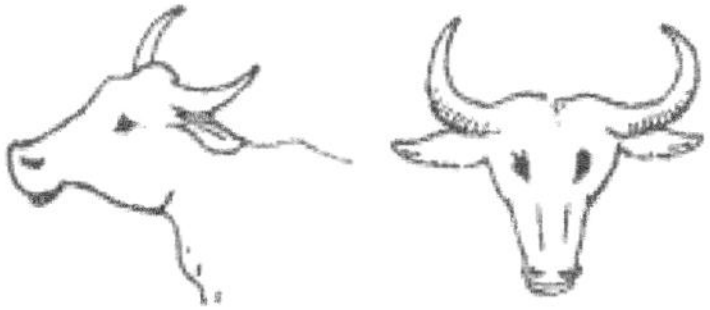

Sheep, goats, calves: 3 cc's (3 mls)

CAUTION!

Precautions:
- Do not give the vaccine if the animal has a fever.
- Shake the bottle well before use.
- Protect medicine from sunlight and heat.

How many times do we give it?
- Give entire dose at one time.
- Repeat dose as required by manufacturer.

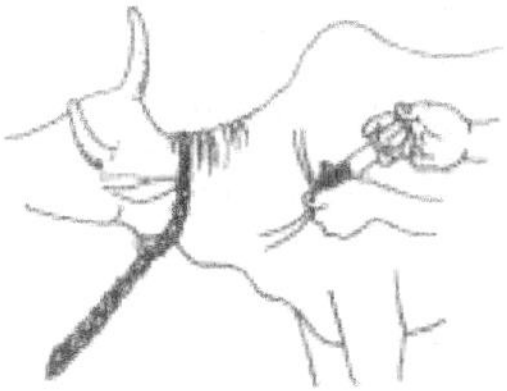

How do we give it?
Inject under the skin *(subcutaneously)*

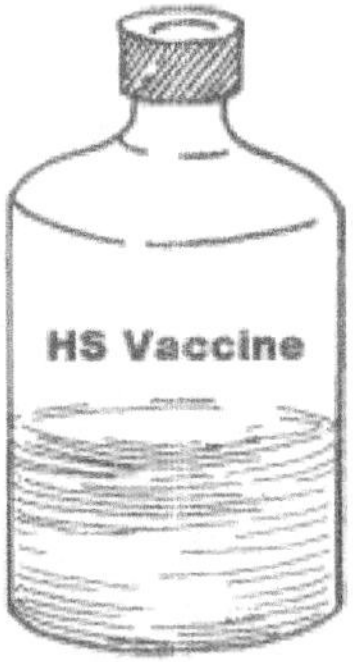

What forms of the medicine are available?
- Broth vaccine is a milky, tan-coloured suspension in 250 ml plastic bottles.
- Oil adjuvant vaccine is also available in some places, but the dose is different. Check with merchant for the correct dose when you buy the vaccine.

FMD Vaccine

When do we use it?
For **prevention** of Foot and Mouth Disease *(khoret)*

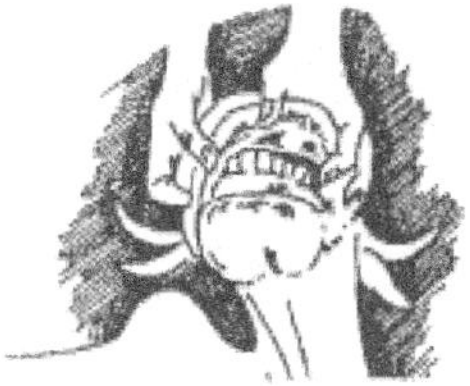

Dose:

Cattle, buffaloes, pigs, and calves:

10 cc's (10 mls)

Sheep and goats: 5 cc's (5 mls)

Precautions:

- Do not give to a pregnant animal during the first and last month of pregnancy.
- Do not give to an animal suffering from FMD.
- Vaccine needs to be kept cold (kept on ice).
- Shake the bottle well before use.

How many times do we give it?

For calves:

- First dose at 6 weeks of age.
- Second dose at 2 to 5 months of age.
- Third dose at 8 to 9 months of age.

For adult animals:

- Give a second dose three months after the first dose.
- Repeat the vaccine dose every 6 months.

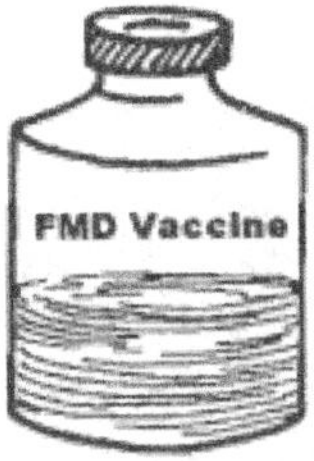

How do we give it?

Give the medicine under the skin *(subcutaneously).*

What forms of the medicine are available?

Milky suspension in 10 ml and 100 ml plastic bottles.

Note: It is advisable to follow the manufacturer's instructions.

F

Eye washing Solution
(2% Boric Acid Solution)

When do we use it?
- For washing the eye to remove discharge.
- For washing the eye to remove infection or foreign material.

What do we need to make it?
- Clean water (about ½ litre)
- Boric acid powder
- teaspoon
- Mixing bowl or glass

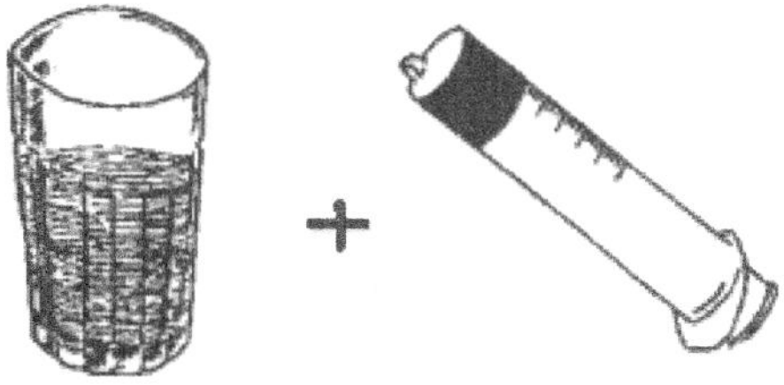

How do we use it?
- Place ½ litre of warm water in a clean bowl or glass.
- Add 1 teaspoonful of powder
- Mix solution until dissolved.
- Take up solution into a clean syringe (without the needle attached).
- Flush eyes until they are clean.

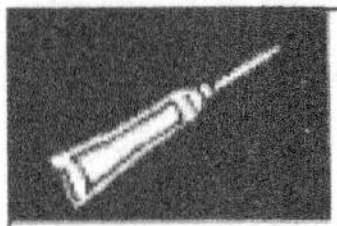

Oxytetracycline Ointment
Terramycin® Eye Ointment

F

When do we use it?

- For the treatment of infected eyes
- Conjunctivitis
- Corneal injury

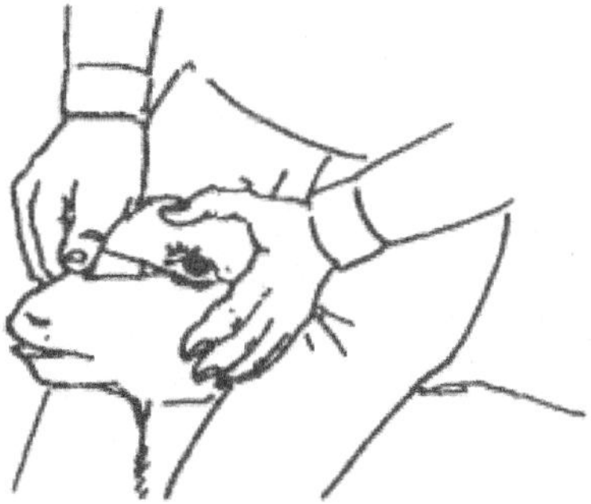

Dose:
Apply enough to cover the eye.

Precautions:

- Do not contaminate the tip of the tube. Keep it clean.
- When applying the ointment, the tube should not touch the eye or eyelid.

How many times do we use it?

2 to 3 times per day until the eye infection is gone.

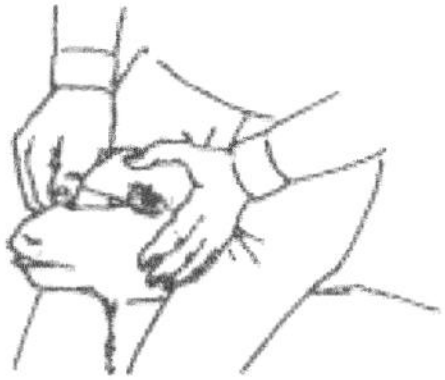

How do we use it?

- Open the eyelid gently with one hand. You may need a helper to hold the head.
- Hold the tube in the other hand and apply the ointment by gently squeezing the tube

Terramycin Ointment

What forms of the medicine are available?

Yellow ointment in small, 3 gram tubes

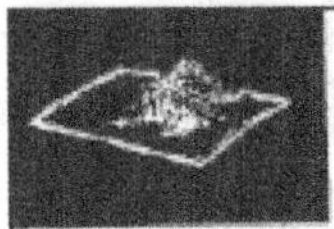

F

Alum

When do we use it?

- To treat eye infections
- To flush wounds
- To clean ulcers in the mouth

Dose:

- For eye infections, apply one pinch of alum powder into the eye.
- For flushing of wounds and mouth ulcers, mix 5 grams of alum in 1 liter of water.

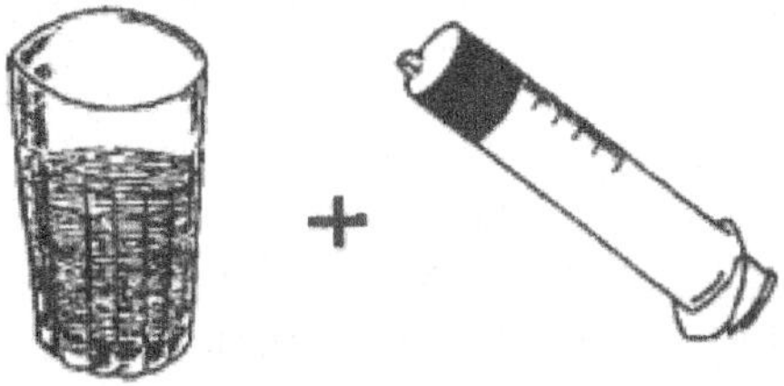

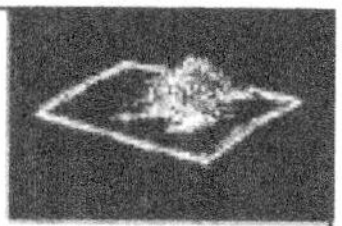

Precautions:

- If the alum powder is coarse, grind it into a fine powder before using in the eye.
- Always use clean hands and utensils when treating the eye.

How many times do we give it?

2 to 3 times per day for 4 or 5 days

Alum powder

How do we use it?

- To treat the eye, place a pinch of alum on a clean paper and blow it into to eye.
- Use a clean, large syringe (without the needle attached) to flush the alum solution into the mouth or into the wound.

What forms of the medicine are available?

- White crystals, like sugar or fine powder
- Commercial packet of 450 grams.

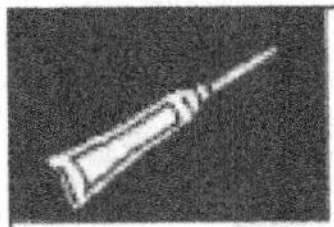

Wide Spectrum Herbal Skin Ointment
Himax®, Herbex®, Charmil®

G

When do we use it?

- To treat all wounds and other injuries
- To treat mange and ringworm
- To treat foot lesions in FMD

Dose:

Apply enough ointment to cover the wound.

CAUTION!

Precautions:

- Always dry the wound before applying the medicine.
- Wash your hands with soap and water after using it.

How many times do we use it?
Once daily until healed

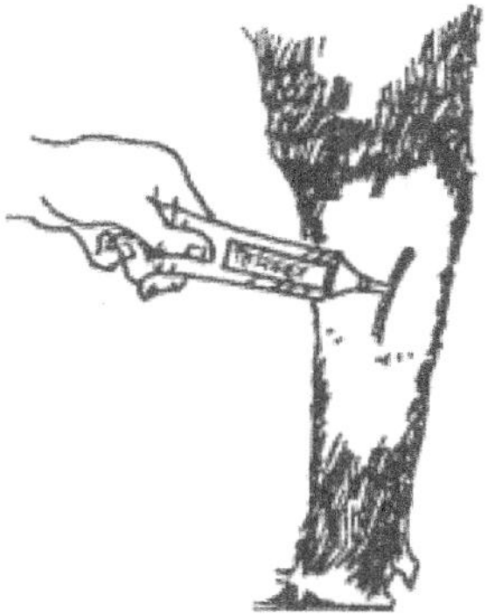

How do we use it?

- Clean the wound with antiseptic solution.
- Dry it with a clean cloth.
- Apply the medicine.

Himax

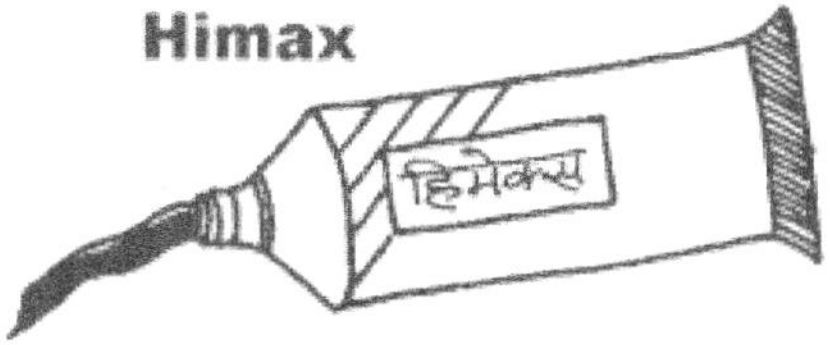

What forms of the medicine are available?

- Brown ointment in 25 gram and 50 gram tubes.
- 1 kg tins

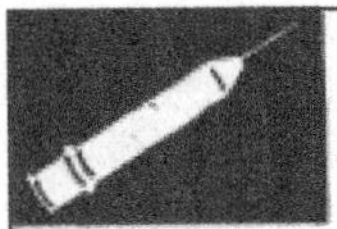

Chlorpheniramine, Avil®, Cadistin®, Cural®

G

When do we use it?

- For the treatment of plant poisoning.
- To treat reactions to medication or allergy.
- To relieve itching from insect bites.

Dose:

Cattle, buffaloes, and horses: 10 mls

Sheep and goats: 2 mls

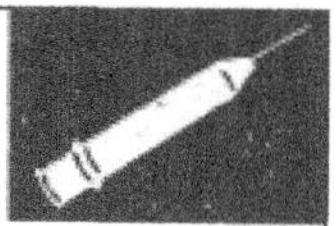

Precautions:
- Must be used immediately after diagnosis of poisoning if it is to be effective.
- There are many different trade names for this medicine. Make sure you have the correct medicine.

How many times do we give it?
- Give entire dose at one time.
- May be repeated every 6 -12 hours as needed.

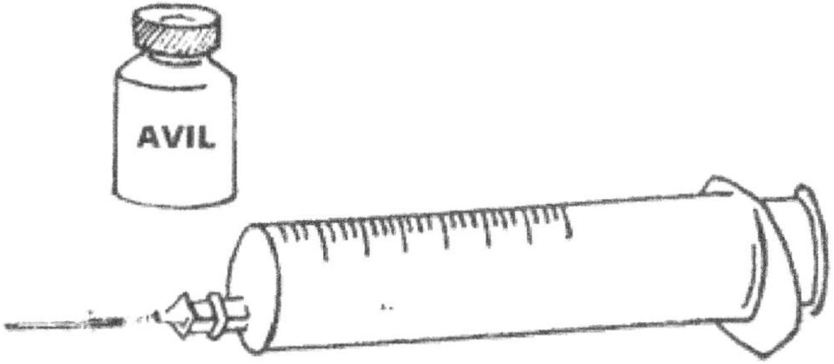

How do we give it?
Inject in the muscle (intramuscularly).

What forms of the medicine are available?
Clear liquid in 10 ml and 30 ml glass vials.

G

Sodium Thiosulphate

How do we use it?
For the treatment of plant poisoning. It works by destroying the poison that the animal has eaten.

Dose:

Sheep and Goats:	10 grams	
Calves:	15 grams	
Cattle and Buffaloes:	30 grams	

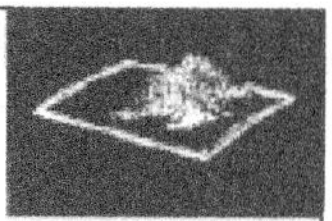

Precautions:

Give immediately as soon as plant poisoning is suspected.

How many times do we give it?

- Give one dose every hour until the animal has recovered.
- Give no more than 3 doses maximum.

How do we give it?

Mix with a bucket of warm water and give with a drenching bottle or tube.

What forms of the medicine are available?

- Coarse, white crystals like sugar
- Commercial 450 gram packets
- Commercial 500 gram and 1 kg packets are also available.

Herbal Uterine Cleanser & Restorative Exapar®

When do we use it?

- For the treatment of retained placenta
- To help expel a dead fetus
- For the prevention of a retained placenta

Dose:

Cattle, buffaloes, and horses: 50 ml

Sheep and goats: 20 ml

Pigs: 5 ml

CAUTION!

Precautions:

To prevent a retained placenta it should be given immediately after delivery of the calf or kid.

How many times do we give it?

- Give 2 doses (morning and night) on the first day, then 1 dose daily.
- Treat for 3 to 5 days or until the placenta comes out.

How do we give it?

- Give with a drenching bottle or tube.
- Mix with sweet feed for horses and pigs.

What forms of the medicine are available?

Brown liquid in 250 ml and 500 ml bottles.

Vitamin B, Beecom-L®, Belamyl®, Bivinal® Livogen®, Conciplex®, Liver-Jeet®

G

When do we use it?

- Inappetence
- Anemia (thin blood)
- Use with dewormer for thin and weak animals.

Dose:

Sheep, goats, and calves: 3 to 5 mls

Cattle, buffaloes, and horses: 10 mls

Note: These doses are for the injectable form. Doses for the oral form are given on the medicine bottle.

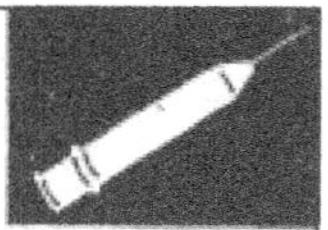

Precautions:

- If medicine has changed color it means that it is spoiled. It should not be used.
- There are many trade names of this medicine available. Make sure you have the right one.

How many times do we give it?

Give once a day for 3 days in a row.

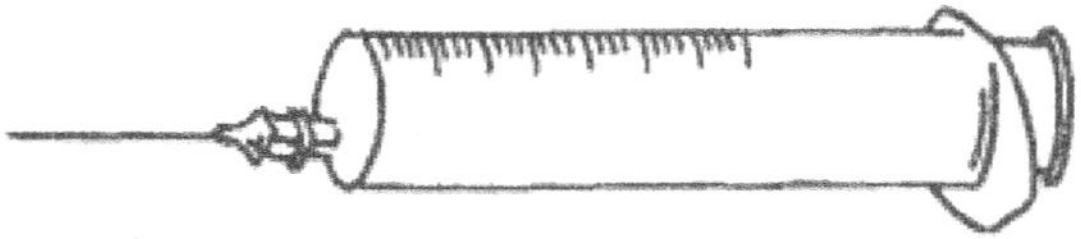

How do we give it?

- Inject in the muscle (intramuscular IM)
- For poultry, it can be mixed into the drinking water. Fresh water should be mixed up each day.

What forms are available?

- Brown liquid in 10 ml and 30 ml glass vials for injection
- Oral liquid to be mixed with drinking water. Many sizes of bottles are available.

Salt-Mineral Block

When do we use it?

- To prevent inappetence
- To prevent anestrus and other breeding problems

Note : If the animal has an appetite for dirt, cement, etc. this might indicate a mineral deficiency.

Precautions:

If using a salt-mineral block, it is not necessary to give extra salt in the diet.

How many times do we give it?

- Allow animal access to the block each day in its stall.
- When the block is gone, give the animal a new one.

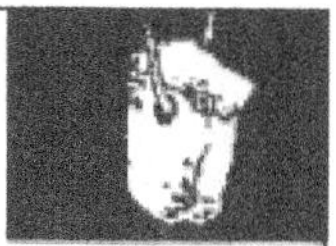

How do we use it?

- Should be hung in the stall near the animal.
- You may need to feed a little by hand at first so the animal gets used to the block.

Note: Hang the mineral block away from water or leaks.

What forms of the medicine are available?

- 2 kg blocks are available commercially.
- You can make a pretty good salt-mineral block yourself.

To make a salt-mineral block, collect the following ingredients:

- ½ kg of red soil
- ½ kg salt
- 5 egg shells
- One handful of flour

Mix these together with a little water to form a paste. Form the paste into a block and make a hole in one end so you can hang it with a string. Place in the sun to dry.

Dettol® Liquid

G

When do we use it?

- For cleaning of wounds
- For relief from insect bites
- To disinfect instruments and utensils

Dose :
Use as much as needed.

CAUTION!

Precautions:
Do not use this medicine in eyes.

How many times do we use it?
As often as needed.

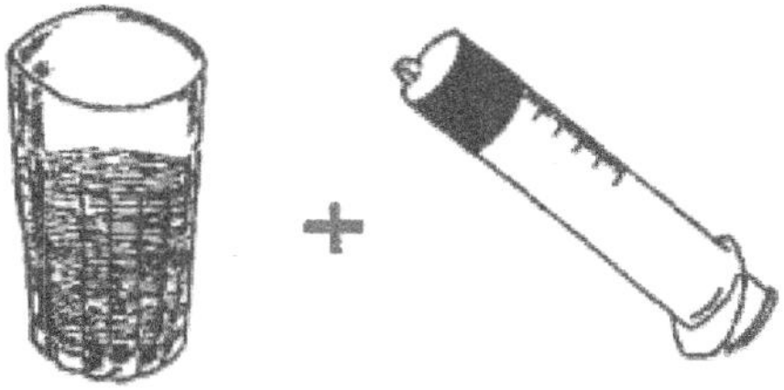

How do we use it?

- Mix 1 part Dettol® with 40 parts clean water for cleaning of wounds and disinfecting instruments.
- You can also use it without mixing with water for insect bites.

What forms of the medicine are available?
Brown liquid in 50 ml, 200 ml, 500 ml, and 1 litre glass bottles

Dettol® Soap

When do we use it?

- For cleaning of wounds
- For cleaning the animal in cases of dystocia and retained placenta
- For cleaning our hands after handling infected materials

Dose:

Use as much as needed.

How many times do we use it?

- Wounds should be thoroughly cleaned before applying any medicine or bandage.
- No foreign material should remain in the wound.

How do we use it?

Mix with clean water.

What forms of the medicine are available?

75 gram and 150 gram soap cakes.

Savlon® Liquid

When do we use it?

- For cleaning of wounds
- To treat mange and ringworm
- To disinfect syringes, needles, and blades.

Dose:

Use as much as needed.

CAUTION!

Precautions:

- Do not leave instruments in Savlon® for more than a few hours. They will become rusty.
- Do not use in eyes.

How many times do we use it?
As often as needed.

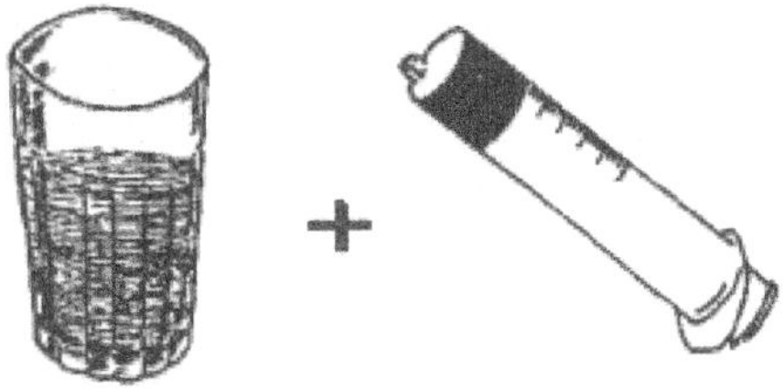

How do we use it?

- For cleaning of wounds and treatment of ringworm and mange, mix one part Savlon® with 10 parts of clean water.
- For disinfection of syringes, needles, and blades, mix 1 part Savlon® with 20 parts water.

What forms of the medicine are available?
Yellow-colored liquid in 100 ml and 1 litre bottles

Pendistrin-SH, Vetclox®, Tilox®, Floclox®

When do we use it?
For the treatment of mastitis

Dose:

- Use 1 tube to treat one quarter of the udder.
- Give entire dose at one time.

Precautions:

- Make sure your hands and the udder and teat are clean before applying medicine.
- Remove all the milk from the quarter before giving medicine.
- Do not drink the milk from a treated animal for 3 days after the last treatment.

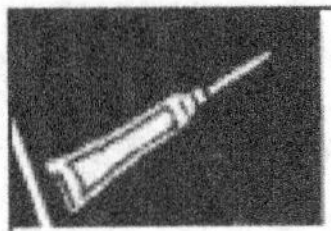

How many times do we use it?

- Medicine should be used morning and night after milking.
- Use twice daily until the animal is well.

How do we use It?

- Wash the udder and teats with soap and water.
- Remove all the milk from the affected quarter.
- Remove the cap from the tube and gently insert it into the teat opening. *Do not get the tube dirty.*
- Gently inject all the medicine into the udder.
- Pinch the end of the teat and massage the medicine up the teat and into the udder.
- Massage the udder to distribute the medicine.

What forms of the medicine are available?

Many kinds of ointments in plastic and metal tubes

Note: Not all mastitis treatment tubes have the same kind of medicine in them, but since they all work the same way, they are included together in this section for convenience.

G

Method for Measuring Powder Medicine

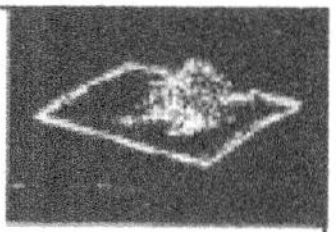

Medicine	Tobacco Can (Full Level)	Tobacco Can Lid (Full Level)
Panacur	6 grams	4 grams
Battisa, Replanta	10 grams	3 grams
Neblon, Afron	10 grams	3 grams
Magnesium sulphate	20 grams	
Sodium thiosulphate	20 grams	8 grams
Sulphur	15 grams	
Amprolium	10 grams	
Stonil	10 grams	
Caflon	g grams	
Timpol	8 grams	

Made in the USA
Monee, IL
05 November 2026

67975446R00203